FROM ATOMS TO QUARKS

FROM ATOMS TO QUARKS

An Introduction to the Strange World of Particle Physics

James S. Trefil

CHARLES SCRIBNER'S SONS • NEW YORK

Library of Congress Cataloging in Publication Data
Trefil, J. S.
From atoms to quarks.
1. Nuclear physics. 2. Particles (Nuclear physics).
I. Title.
QC777.T73 539.7 80-11093
ISBN 0-684-16484-1
ISBN 0-684-17460-X

Published simultaneously in Canada
by Collier Macmillan Canada, Inc.

1 3 5 7 9 11 13 15 17 19 H/C 20 18 16 14 12 10 8 6 4 2
5 7 9 11 13 15 17 19 F/P 20 18 16 14 12 10 8 6

Printed in the United States of America

To my mother and to Flora

Acknowledgments

There are many people who helped me with the preparation of this book. I would like to thank especially my son Jim for serving as a sounding board for early drafts of the manuscript, my son Stefan for help in preparing the illustrations, and Mrs. Nancy Lane for her assistance in getting the manuscript into finished form.

James S. Trefil
Red Lodge, Montana
July 27, 1979

Contents

A Brief Introduction

IT HAS become a truism that the most exciting frontiers in the physical sciences are the study of the very small and the very large. There are many excellent books available that deal with the latter subject, but very few that deal with the former. The past thirty years have seen a remarkable breakthrough in our understanding of the atomic nucleus and the particles that exist therein, and in the last five this trend has accelerated. We are now at a point where there are tantalizing glimpses of possible resolutions to one of the oldest questions men have asked: the question of what the world is made of. In this book, my intention is to take the reader through the sequence of discoveries that led from the atom to the nucleus to the elementary particle and, finally, to the quark. Whether this new particle (which has yet to be seen in the laboratory) will ultimately prove to be the truly "elementary" building block of all matter remains to be seen, but the reader who follows the story presented here will finish with a pretty good idea of where we now stand in the quest.

FROM ATOMS TO QUARKS

CHAPTER I

The Quest for Ultimate Simplicity

Gaily bedight,
A gallant knight.
In sunshine and in shadow,
Had journeyed long,
Singing a song,
In search of Eldorado.

—EDGAR ALLAN POE,
"Eldorado"

Introduction

THERE is something in man that makes him want to know. He looks at the world and asks questions. What is it made of? Why is it there? How does it work? Some of this yearning to know is expressed in religion, some in art, and some in philosophy. But for those who share the general background usually called Western culture, the yearning is most completely expressed in the institution called science.

Physics is the science that is concerned with matter and motion, and it is to physics that some of the most interesting of these basic questions apply. In particular, the questions about how the material world is put together and what it is made of have long occupied the thoughts of men who would now be called physicists. Although these men lived thousands of miles and thousands of years apart, a single concept runs through their ideas like a unifying thread. Perhaps concept is the wrong word. Perhaps we should call it a belief, a hope, a dream. From the time men first started asking scientific questions, they have assumed that when they knew enough to answer the questions fully, they would find that the world was really a simple, uncomplicated sort of place.

When you consider the overwhelming complexity that we see around us, you have to appreciate what a tremendous leap of faith this was. Yet in about 585 B.C., Thales of Miletus, the Ionian Greek who is called the first scientist, noticed that matter comes in three forms—liquid, solid, and gas—and suggested that the entire world might be nothing more than water (which is also seen as liquid, ice, and steam). From this beginning his students developed the familiar Greek system in which everything known is some combination of just four elements—earth, fire, air, and water. This theory, containing as it does four basic constituents, is the first serious attempt to find a scheme that is both simple and explains the observed complexity of the world. If we follow modern usage, we would call earth, fire, air, and water elementary or fundamental, and if we believed that they came in small pieces, we might even call them elementary particles.

In the late fifth century B.C., the philosophers Leucippus and Democritus, trying to resolve the conflict between the observed complexity and transience of the physical world with the Greek idea that Truth must be eternal and unchanging, suggested that matter might actually be composed of small particles. They called these particles atoms (literally, that which cannot be split), and pointed out that while atoms could be unchanging, the relationship between atoms could change. In this way, both of the seemingly contradictory observations could be satisfied in a single philosophical system. Although this idea appeals to us a great deal in hindsight, it enjoyed little currency among Greek and Roman philosophers (except for a brief revival by the Epicureans), and it remained something of a philosophical sidelight until modern times.

The first statement of the modern atomic theory was put together in the early nineteenth century, when John Dalton, an English chemist, published his two-volume book called *A New System of Chemical Philosophy* (1808). In this book Dalton pointed out that many of the laws of chemistry that were known in his time could be explained easily if it were assumed that to each chemical element there corresponded an atom of matter. The atom for hydrogen was different from the atom for sulphur, of course, but the basic idea was clear. Every substance in the world must be made up of different combinations of a few different kinds of atoms, so that in this picture it is the atoms themselves that would be the fundamental building blocks of matter. Consequently, it is to the atoms that we would award the title "elementary particle." And even if today we know of over 100 chemical elements (as opposed to 26 in Dalton's time), it can still be argued that the modern atomic theory does impose a kind of simplicity on nature.

To Dalton, as for Democritus, the atoms were indivisible. Indeed, they were represented as featureless spheres in most of his books. Consequently, there was little in the atomic theory to prepare the scientific mind for a rather astounding series of discoveries that took place between 1890 and 1920.

The Discovery of the Electron

By the middle of the nineteenth century, physicists were experimenting with a new phenomenon that was eventually to show that the simple atomic picture of matter had to be drastically changed. By that time it had become possible to produce an experimental apparatus similar to that shown in Illustration 1, in which a glass tube had a metal plate (called an *anode*) at one end, a wire capable of carrying electrical current (called a *cathode*) at the other, and a pretty good vacuum in the tube itself. By passing electrical currents through the cathode, physicists hoped to study the electrical properties of the rarefied gas in the tube.

When current was introduced in the cathode of such a device and the anode was held at a high positive voltage, a strange phenomenon was seen. A thin line of glowing gas formed near the cathode and extended toward the anode. From an analysis of the light emitted by the tube, it was clear that this thin line was made up of residual gas that had been heated when something passed through it. The unknown something was called *cathode rays,* and the consuming question then centered around their identity.

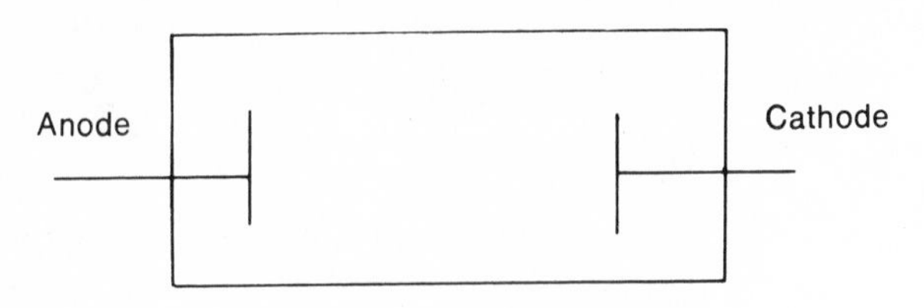

1. Experimental apparatus similar to the one used in the discovery of cathode rays.

At that time, physicists knew that an object that carried an electric charge could be affected by two kinds of forces. If it were brought near the pole of a battery, it would be either attracted or repelled, depending on whether the positive or negative pole of the battery was involved. In technical language, it was said that such a particle was acted on by an electrical force.

In a similar way, if an electrically charged object was moving and a magnet were brought near its path, the charged object would change direction. This was taken as evidence that a magnetic force existed in addition to the electrical force. These two are sometimes referred to as *electromagnetic* forces.

The electric and magnetic forces will act on any charged object, but will not act on something that carries no charge, such as light. Thus, the debate on cathode rays eventually came down to the question of whether they possessed an electrical charge.

By the early 1880s, there were two definite schools of thought on this topic. In Germany the general feeling was that the cathode rays were a new type of radiation, something like light. To support this view, they pointed out that light was capable of transmitting energy in the form of heat. Isn't sunlight, after all, the basic source of warmth on the earth? In England it was felt that the cathode rays had to be particles of some type. The primary reason for this belief was that it was known that the glowing line that marked the passage of the cathode rays could be deflected and moved by bringing magnets near the apparatus, and this could only mean that the cathode rays were affected by the magnetic force. Since this had never been observed for light, the thinking went, the rays must be something else. The primary theory in England at the time was that atoms in the gas, upon striking the cathode, somehow became negatively charged and were then pulled through the gas toward the anode.

In 1897, J. J. Thomson, a young English physicist, performed a series of experiments that seemed to settle the matter once and for all. He reasoned that if the cathode rays were really particles, they would not only be affected by magnets but by large electrical charges as well. He knew that by measuring the amount by which a particle of known velocity is deflected by a magnetic field, it would be possible to determine the ratio of the particle's charge to its mass. (We will talk about the details of how this is done in Chapter VI.) If the charge of the cathode ray were called e and the mass m, then the charge to mass ratio would be e/m. The problem was that no one knew how to find the velocity of the cathode rays. If a way could be found to measure it, scientists would know whether the cathode rays were really radiation, because if they were, their velocity would be the same as that of light. As an added benefit, scientists would also know the value of e/m of the particle.

The apparatus that Thomson used is shown below (Illus. 2). The cathode rays were directed across a region between two charged plates, and in this region there was also a magnetic field. What he did was to adjust the charge (or, equivalently, the voltage) on the plates until it exactly canceled the deflecting effects of the magnetic field. In other words, if the magnetic field would cause the cathode rays to move downward, it would mean that the plates were charged in such a way as to move the beam back up by an equivalent amount. By measuring the electrical field that will exactly cancel the deflection due to the magnetic field, the velocity of the particles can be found. A more detailed description of this process will be given in Chapter VII.

The important result that Thomson obtained was that the velocity of the cathode rays was about 3×10^7 kilometers/second, which is about a tenth of the velocity of light. Clearly, the cathode rays were particles. From the fact that they were attracted toward a positively charged anode, he could conclude that the particles carried a negative electrical charge. The particle is now known as the *electron,* and its mass is 9.1×10^{-28} gram—a very small mass.

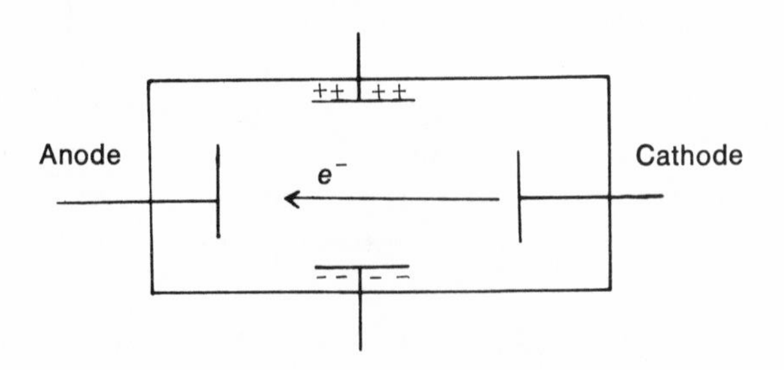

2. A sketch of the apparatus used by Thomson to detect the electron. As the electron moves to the left in the tube, it is attracted upward by the charged electrical plates. The effect of the plates is exactly canceled by a magnet *(not shown)* with its north pole below the plane of the paper and its south pole above.

Once it was properly identified and labeled, it was realized that the electron is a very important particle indeed. Every electrical current, whether it is a man-made circuit or is in a nerve in the body, is simply the flow of electrons. For example, it takes about 6,000,000,000,000,000,000 electrons *per second* to keep a 100-watt light bulb going. Moving electrons are, in fact, the "electrical fluid" that Benjamin Franklin first postulated to explain electricity in his time.

From the point of view of the search for simplicity, however, the discovery of the electron was rather ominous. After all, the only place the particle could come from was the interior of some "indivisible" atom. The existence of a negatively charged particle that can be taken from the atom implied that there must be a positively charged segment left behind, and this, in turn, implied that the atom must have structure. If this was so, then there must be a type of matter more fundamental than the atom. The electron was the first example of matter from this deeper level.

The Photon

LIGHT has been known to be a wave since the eighteenth century, but the discovery that its origins are tied to electrical charges and that it is only one example of electrically generated waves was one of the great triumphs of nineteenth century physics. From the discovery have flowed some of the most important artifacts of our culture—radio, television, radar, and microwave technology, to name but a few. In this section, I shall talk about the idea of light as a wave and then show how some developments early in the twentieth century suggested another way of looking at it.

A classical example of a wave is surf coming into a beach. The wave is characterized by three numbers—the wavelength, or distance between crests of the wave; the frequency, or the number of crests that go past each second; and the amplitude, or maximum height of the wave. These quantities, in general, are independent of each other. In fluids it is possible, in principle, to have any combination of wavelength, frequency, and amplitude in a wave.

If you have ever seen surf coming into a rocky coast, you realize that waves can carry tremendous amounts of energy. The energy carried by a wave always depends on the amplitude, and for some waves it can depend on the other two parameters as well. Thus, we could just as well characterize a wave by frequency, wavelength, and energy, as by frequency, wavelength, and amplitude. This is the most useful way to think of light waves.

Finally, we note that if the frequency is the number of crests passing a point per second and the wavelength is the distance between crests, the speed of the wave must be $V = \lambda\nu$, where we have followed the usual convention and represented the wavelength by the Greek letter λ (lambda) and the frequency by the Greek letter ν (nu). Remember that this is the velocity of the wave and not of the water. As a particular wave moves by a point, the water moves up and down—it does not move along with the wave.

The identification of light with this phenomenon rests on arguments too detailed to go into here, but essentially it depends on the fact that light seems to behave in ways that are very similar to water waves. The discovery about light alluded to above had to do with how the waves are generated. It turns out that whenever electrically charged objects are accelerated, a wave is generated that moves outward from the objects in an ever expanding sphere. It is a three-dimensional analog of the circular wave that moves away from the point where a rock is thrown into a still body of water. Because the wave moves away from the charged object, it is called *radiation,* and because it is electromagnetic in origin, it is called *electromagnetic radiation.* In modern terms, the phenomena that correspond to the displacement of the water in surf are oscillating electrical and magnetic fields through moving space.

In classical physics, there is no reason why these electromagnetic waves cannot have any combination of wavelength and energy. Theory requires, however, that all the waves move at the same velocity. This velocity is so important that it is given a special letter—c—and it has the experimentally determined value

$$c = 3 \times 10^8 \text{ m/sec} = 186{,}000 \text{ mi/sec}$$

This means that if we know either the frequency or wavelength of a particular radiation, we can determine the other from the relation $c = \lambda\nu$. The most common examples of radiation at different wavelengths are the colors that we can see with the unaided eye. Each different color corresponds to a different wavelength and frequency of light. Red has the longest wavelength and lowest frequency, while violet has the shortest wavelength and highest frequency.

In more quantitative terms, visible light corresponds to electromagnetic radiation, for which the wavelength is between 3.8×10^{-5} centimeters (violet) and 7.8×10^{-5} centimeters (red). Clearly, this rather narrow band of wavelengths does not exhaust the possibilities for electromagnetic radiation. Even subject to the constraint in the equation, there is a continuum of possible wavelengths. Some typical familiar electromagnetic waves are listed in the table, together with the names usually assigned to each wavelength interval.

ν	λ (cm)	
10^6	3×10^4	AM radio
10^7	3×10^3	shortwave radio
10^8	3×10^2	FM radio and TV
10^{10}	3	microwave
3×10^{13}	10^{-3}	infrared
3×10^{16}	10^{-6}	ultraviolet
3×10^{18}	10^{-8}	X ray
3×10^{20}	10^{-10}	gamma ray

Once the infinite variety of possible electromagnetic radiation was realized, some problems arose. The central problem seemed to be that if the radiation of a given frequency had an energy independent of the frequency, the principles of classical physics seemed to predict that the universe would be full of high-energy radiation, which is not the case. To get around this problem, the German physicist Max Planck suggested in 1900 that these problems could be resolved if the energy of radiation emitted by an atom could not be just any number, but had to be an integral multiple of a basic energy given by $E = h\nu$. In this equation, E is the energy of the wave, ν its frequency, and h is a constant, now called *Planck's constant.* We shall encounter this quantity again, since it plays an important role in all interactions on the atomic level. For our present discussion, the important point is that this equation seems to suggest that light is emitted in little bundles rather than in some kind of continuous spectrum. Planck called these little bundles *quanta,* but he never really accepted the idea that light was not an ordinary classical wave.

Albert Einstein, however, saw that Planck's quantum postulate could have far-reaching consequences if it were taken to its logical conclusion. For example, there was an experimental phenomenon known as the photoelectric effect, in which it was observed that when light was allowed to shine on one side of certain metals, electrons were emitted from the other side. This phenomenon could be understood in the classical wave theory—the electromagnetic waves would move electrons out of the metal in much the same way as the surf brings driftwood to a beach. The problem was that it would take a classical wave a long time to work an electron out of an atom, and the experimental fact was that electrons were emitted from the metal as soon as the light was turned on.

In 1905 Einstein suggested that if light were really composed of particlelike bundles rather than being a classical wave, the above experimental fact could be understood. If Einstein's theory was true, then the interaction between light and electrons would be analagous to a collision between billiard balls, and the electrons would come flying out immediately after the collision. For this work, Einstein was awarded the Nobel Prize in 1921. (The theory of relativity was then too "risky" for the conservative Nobel committee to honor.)

The particles that make up light and other electromagnetic radiation are now called *photons.* They can be thought of as the constituents of radiation, and in this sense they would have to be included in any list of elementary particles that are supposed to give a simple explanation of matter.

The Nucleus and the Proton

In the early 1900s, a rather extraordinary figure appeared on the stage of world science. The son of a New Zealand wheelwright, Ernest Rutherford became the leading figure in the exploration of atomic structure and is largely responsible for the picture of the atom that we have today.

At this time, a great deal of work had been done on the chemistry of radioactive elements. It was known, for example, that there were three types of radiation given off by these elements, and one of the important tasks was to identify them. They were called α (alpha), β (beta), and γ (gamma) radiation, respectively, and each had different properties. In modern terms, the beta rays are electrons and the gamma rays are high-energy photons. Working at McGill University in Montreal, Rutherford was able to show that the mysterious alpha rays were, in modern terms, the nuclei of helium atoms. He was able to show this by performing experiments in which small samples of naturally radioactive elements that were known to emit alpha radiation were placed near sealed, evacuated tubes. After a period of time, sensitive chemical analysis showed the presence of helium in the tube where none had existed before. Since only the alpha radiation could have entered the tube, the connection between these rays and helium was established. For this experiment, and for other work in sorting out the behavior of radioactive elements, Rutherford was awarded the Nobel Prize in chemistry in 1908.

Contrary to the usual rule, Rutherford actually did his most important work *after* he received the Nobel Prize rather than before. When he went to Manchester University in England, he continued his experiments with alpha particles. One of the "hot" topics in those days was the study of the way in which these particles passed through thin metal foils. The idea was that you could learn something about an atom by seeing how the alpha particles bounced off of it. Almost as an afterthought, he suggested to some co-workers in 1911 that it might be interesting to see if any particles were ever bounced back (i.e., suffered a change of direction of 180°) when they hit a nucleus.

According to the ideas current at the time, the atom was a large, diffuse, positively charged material in which electrons were embedded like raisins in a bun. Such an atom could not deflect an alpha particle, anymore than a cloud of gossamer could deflect a bullet. When Rutherford's experiment was carried out, however, a surprisingly large number of alphas—perhaps one in a thousand—were scattered through angles close to 180°. This could only be understood if it was assumed that the atom, far from having its mass smeared out over a fairly large area, actually had virtually all of its mass concentrated in a small spot in the center. This concentration of mass Rutherford called the *nucleus* of the atom. Since the nucleus repelled the positively charged alpha particles, it must also be positively charged.

The picture of the atom that Rutherford evolved would be familiar to us (Illus. 3). The atom consisted of a small, dense, positively charged

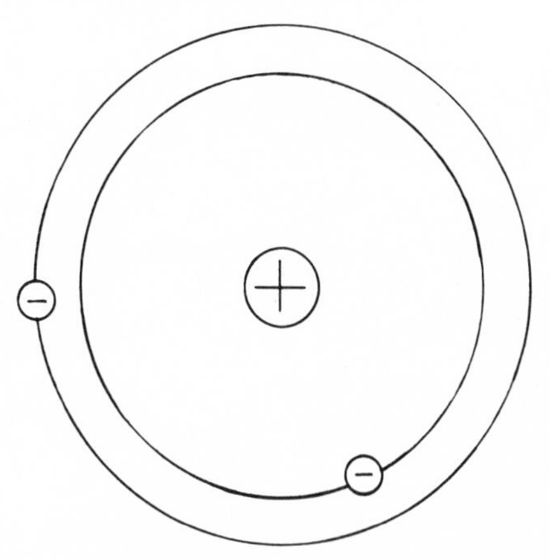

3. The atom as Rutherford envisioned it.

nucleus in which most of the mass resided and around which the electrons orbited. The analogy between the nuclear atom and the solar system was remarked upon by many observers at the time, and served as the plot for innumerable science fiction stories during the 1930s and 1940s. In that sense, it has become part of the accepted folklore of modern culture.

Once the existence of the nucleus became established, scientists could begin to ask about its composition. Was it a uniform blob of positively charged matter analagous to Dalton's atom, or were there constituents inside, waiting to be discovered? In 1919, using techniques similar to those that had allowed him to identify the alpha particle, Rutherford was able to show that a certain particle emitted from nuclear collisions of alphas with nuclei was itself the nucleus of hydrogen. Since hydrogen is the lightest atom, it follows that its nucleus must play some special role in nature. Rutherford recognized this fact by naming the new particle the *proton* (the first one).

The proton is a particle that has a positive electrical charge. The magnitude of this charge is precisely equal to the magnitude of the charge on the electron, although the signs of the charges of the two particles are opposite. The hydrogen atom must therefore be a system similar to the one shown in Illustration 4. A single electron orbits a single proton, making hydrogen the simplest possible atom that can be imagined. Since protons are found in the debris of nuclear collisions, they must exist in heavier nuclei as well. In fact, each unit of positive

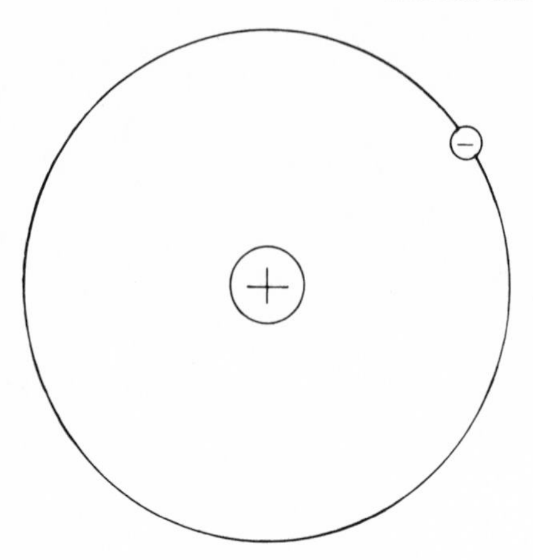

4. The hydrogen atom.

charge in a nucleus is supplied by a proton, so that helium (the next heavier element after hydrogen) must have two protons in its nucleus, lithium three, and so forth. The heaviest natural element, uranium, has a total of ninety-two protons in its nucleus.

As we might expect from the Rutherford experiment, the proton is much heavier than the electron. Its mass is now known to be

$$m_p = 1.67 \times 10^{-24} \text{ g}$$

so that the ratio of the proton to the electron mass is 1,836.

In passing we should note that in 1920 Rutherford suggested that there was probably another constituent to heavier nuclei; that is, an electrically neutral particle about as massive as the proton. He called this hypothetical particle the neutron. He came to this conclusion by noting that most atoms seemed to weigh about twice as much as one would expect them to if one added up the masses of the protons and electrons in them. The neutron and its eventual discovery in 1932 will be discussed in Chapter II.

A Remark About the Bohr Atom

ONCE it was realized that the electrons in an atom move in orbits around the nucleus, physicists found themselves in a quandary. According to the laws that govern charged particles, an orbiting electron should emit light, lose its energy, and eventually fall into the nucleus. Calculations indicated that for the hydrogen atom (made up of one proton and one electron), this process would take less than a second. Since hydrogen has been around since the beginning of the universe, there was clearly something wrong with these arguments. In 1913 a young Danish physicist by the name of Niels Bohr suggested a way out of this dilemma.

The essential difference in Bohr's picture of the atom is that electrons circle the nucleus only at certain well-defined distances. For example, in the hydrogen atom sketched in Illustration 5, the electron can be in an orbit of radius r_2 or r_1, but it cannot be at any orbit between the two. The orbits at radius r_2 and r_1 are known as *allowed orbits.* Why the atom should arrange itself in this way did not become clear until the advent of quantum mechanics in the next decade, but the theoretical difficulties with the classical nuclear atom do not appear in the Bohr atom. In addition, the Bohr atom provides a very simple visualization of the process by which atoms emit light. If an electron happens to be in the orbit at radius r_1, the only way it can move to a lower orbit is to make an instantaneous jump, called a *quantum jump.* For example, the sketch shows the electron making a quantum jump from orbit r_1 to orbit r_2. But these two orbits have different energies, so in order for things to balance, the lost energy must be radiated away from the atom. It is, in fact, carried away by a photon whose energy is equal to the

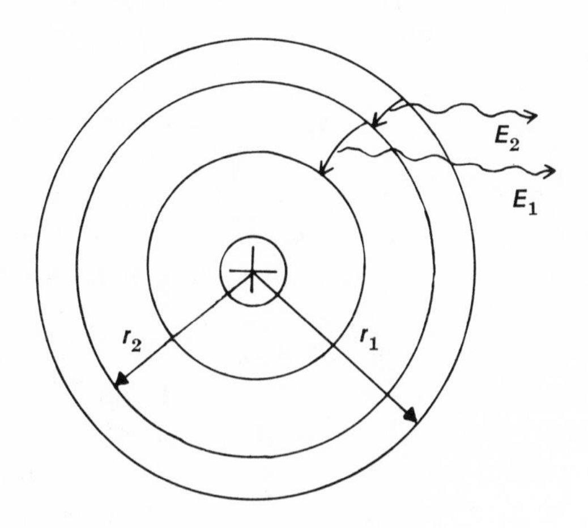

5. The electron's allowed orbits and quantum jumps.

differences in energy between the two orbits. If the energy of the photon is right, we will see the light emitted by the electron transition, and we say that the atom is in a piece of material that is "red hot."

In the orbit r_1 the same process can occur. The electron could jump to the next lower orbit and another photon could be emitted. Alternatively, the electron could jump from the orbit r_2 to the orbit below r_1 and emit a single higher energy photon. The complete working out of the connection between an atom's structure and the light it emits is the domain of atomic physics. For our purposes, the important point to note is that Planck's assumption of the existence of quanta is mirrored in the way electrons are arranged in an atom.

Detection of Charged Particles

IF elementary particles are going to be searched out and studied, it is clear that some way of detecting them has to be found. Thus far, we have glossed over the details of how this is done. Particles that carry an electrical charge are the easiest to detect; hence, they were the first elementary particles found. The reason for this easy detection is that when a charged particle approaches an atom, the electrical force acts between the particle and the electrons in the atom, so that the atom is changed during the passage. In one common occurrence, the particle actually tears an outer electron from the atom, leaving in its wake a free electron and a residual atom that has a net positive charge. An atom that is missing an electron is called an *ion,* and the process described above is called *ionization.*

Alternatively, the passage of the charged particle could simply leave behind an atom in which one or more electrons were lifted to orbits far away from the nucleus. In that case, the electrons would eventually jump back down to lower orbits, and radiation would be emitted. This radiation could be perceived as visible light. A material that emits light when a charged particle enters it, either through the process we have described or a more complicated one, is called a *scintillator.* Many of the early experiments on alpha particles were done by people sitting in darkened rooms watching for the flashes of light that indicated that a particle had struck a sheet of scintillating material, which had, in turn, scintillated.

It has been the detectors that depend on ionization that have played the major role in the exploration of the world of the nucleus. Perhaps the most commonly known detector is the Geiger-Müller counter (usually just called the Geiger counter). This instrument consists of a rarified gas in a long metal tube, as shown in Illustration 6. A wire runs down the center of this tube. By connecting this wire to one side of a voltage source (such as a battery) and the metal walls to another side, it is possible to create a voltage between the wire and the wall of the counter.

Suppose that a charged particle now passes through the chamber. It will leave behind it some free electrons and positive ions, as we discussed above. If the central wire has been connected to the positive pole of the voltage source, then the positive ion will be attracted to the wall and will start to move in that direction, as shown. Similarly, the electron will start to move toward the wire. If the voltage is high enough, the electron will soon be moving fast enough so that it, too, can ionize atoms of the gas and create new electrons that, in turn, will ionize still other atoms and create still more electrons. This process then builds up an *avalanche* in the gas, and when the large numbers of electrons produced in this way strike the central wire, a large current will appear. This signal, suitably amplified and fed through a loudspeaker, produces the familiar "click" that the Geiger counter emits in the presence of radioactive material.

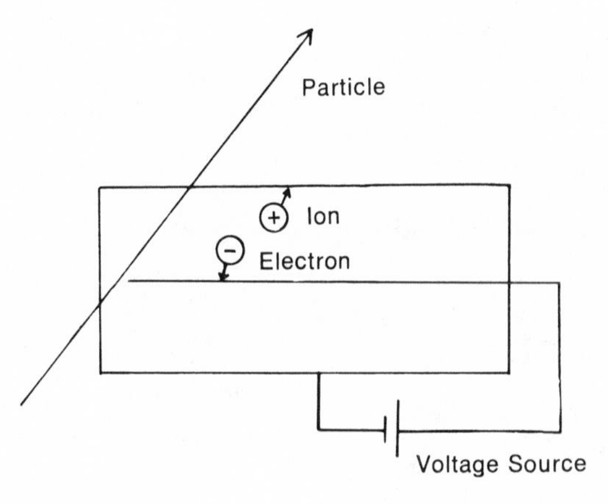

6. Sketch of the Geiger counter.

The principle of the Geiger counter (creating extra electrons by applying a voltage and then detecting the enhanced signal) is used in many different types of counters in modern physics. This short sketch of its operation, however, should give you some idea of the processes involved when a physicist says "A particle was detected. . . ."

The Size of Things

In familiar units, a typical nuclear diameter might be, say, 0.000000000000006 meter, and the speed of light is 300,000,000 meters/second. Rather than go through the bother of writing all of those zeros, physicists use a notation that makes handling both large and small numbers more convenient. I have seen it called scientific notation and powers-of-ten notation, as well as other titles. I see no point in quibbling over the name we use for it; we shall just use it without worrying about what it is called.

The key point is that every number, no matter how large or how small, is written as a number between 1 and 10 times a power of 10. For example, the number 250 would be written 2.5×10^2 and the number 0.003 as 3.0×10^{-3}. A useful mnemonic for this notation is to think of a positive exponent in the power of 10 as a direction to "move the decimal point to the right," and the magnitude of the exponent as saying how many places it should be moved. Similarly, negative powers of 10 imply that the decimal point should be moved to the left. The number 2.5×10^2 would therefore be interpreted as the direction "move the decimal point in the 2.5 two places to the right," while the number 3.0×10^{-3} would mean "move the decimal point three places to the left."

In this notation, the nuclear size noted above would be 6×10^{-15} meters, and the speed of light would be 3×10^8 meters/second.

Arithmetic manipulations in this notation are also quite simple. To multiply two numbers, you multiply the numerals in the usual way and add the powers of 10. For example, to multiply 250 by 250, we write:

$$2.5 \times 10^2 \times 2.5 \times 10^2 = 6.25 \times 10^4 = 62{,}500$$

To multiply 250 by 0.003, we write:

$$2.5 \times 10^2 \times 3 \times 10^{-3} = 7.5 \times 10^{-1} = 0.75$$

The rule for division is that the exponent of the denominator is subtracted from the exponent of the numerator, so that 250 divided by 0.003 would be

$$\frac{2.5 \times 10^2}{3 \times 10^{-3}} = 0.83 \times 10^5 = 8.3 \times 10^4 = 83{,}000$$

while 0.003 divided by 250 would be

$$\frac{3 \times 10^{-3}}{2.5 \times 10^2} = 1.2 \times 10^{-5} = 0.000012$$

As an example of how this notation is used, let us see how long it would take light to cross a typical nucleus. This should be the distance traveled divided by the velocity. Using the numbers above, we find the transit time to be

$$t = \frac{6 \times 10^{-15}}{3 \times 10^8} = 2.0 \times 10^{-23} \text{ sec}$$

This very short time will acquire a great deal of significance later, since we shall see that events inside a nucleus seem to take about this long to happen.

We can use the notation to discuss the size of some of the things we have talked about so far. If we consider a typical nucleus to have 10–20 objects the size of a proton in it, and a typical atom to have 5–10 electrons, then the table below lists some typical sizes. (For reference, oxygen has sixteen proton-sized objects in its nucleus and eight electrons in orbits.) The length 10^{-15} meter is an important one, and is therefore given a special name. It is called the fermi (after Enrico Fermi, whose work we will discuss later).

From the following table an important quantitative fact emerges. While the proton and the nucleus do not differ greatly from each other in size, the nucleus is much smaller than the atom. Rutherford had

QUANTITY	LENGTH (m)
Radius of the proton	8×10^{-16}
Radius of a typical nucleus	3×10^{-15}
Radius of a typical atom	3×10^{-10}

conjectured that the atom was mostly empty space, but with the information in the table we are now in a position to see just how empty it is.

We can start by comparing the volume of the nucleus to the volume of an atom, assuming that both are spheres. The formula for the volume of a sphere of radius R is

$$V = \frac{4}{3} \pi R^3$$

so the volume of the proton is

$$V_P = \frac{4}{3} \pi (8 \times 10^{-16})^3 = 2.1 \times 10^{-45} \text{ m}^3$$

while the volume of the nucleus is

$$V_N = \frac{4}{3} \pi (3 \times 10^{-15})^3 = 1.1 \times 10^{-43} \text{ m}^3$$

and the atom has a volume of

$$V_A = \frac{4}{3} \pi (3 \times 10^{-10})^3 = 1.1 \times 10^{-28} \text{ m}^3$$

All of these are very small volumes, of course, but we can compare them. For example, we can ask what fraction of the volume of a typical nucleus would be taken up by a single proton. From the above numbers,

$$\frac{V_P}{V_N} = \frac{2.1 \times 10^{-45}}{1.1 \times 10^{-43}} = 1.9 \times 10^{-2} = 0.019 = 1.9\%$$

Thus, a single proton is roughly 2 percent of the volume of our typical nucleus. Since this nucleus has 10–20 proton-sized objects in it, we would say that these objects take up an appreciable fraction of the nuclear volume. The nucleus, in other words, is definitely *not* mostly empty space.

Quite a different conclusion results when we compare the size of the nucleus to the size of the atom. In this case,

$$\frac{V_N}{V_A} = \frac{1.1 \times 10^{-43}}{1.1 \times 10^{-28}} = 10^{-15}$$

This means that, expressed as a percentage, the nucleus occupies 0.0000000000001% of the volume of the atom. The rest, except for the very small electrons, is empty space.

We can make this point in another way. Suppose the nucleus were blown up in size until it was a foot across; that is, the size of a large bowling ball or a small watermelon. How big would the atom be? Since our typical atom has a radius 10^5 times larger than the nucleus, the expanded atom would have to have a diameter of 10^5 feet. Since there are 5,280 feet in a mile, this would correspond to

$$\frac{10^5}{5{,}280} = 19 \text{ mi}$$

Hence, if the nucleus of our atom were as large as a bowling ball, the rest of the atom would consist of perhaps ten pea-sized electrons scattered around a sphere 20 miles across with the bowling ball at the center. Imagine putting a bowling ball in the center of a city and then scattering ten peas around through the rest of the city and you will have some idea how empty an atom really is.

CHAPTER II

The Nucleus

. . . the centre cannot hold.

—WILLIAM BUTLER YEATS,
"The Second Coming"

What Keeps It Together?

ONCE we recognize that most of the matter in an atom is packed into a nucleus, our attention is naturally drawn to the way that the nucleus is put together. From Chapter I we know that the nucleus has a positive electrical charge and that it contains protons. One of the basic laws of electricity tells us that while opposite electrical charges must attract each other, similar electrical charges are repelled. This means that in any nucleus other than hydrogen (i.e., in any nucleus containing more than one proton), there will be a force that tends to push the protons apart. If this repulsive force were not countered by some other force, the nucleus of every atom would fly apart. Since they do not, we can conclude that there must be some sort of force acting in the nucleus that tends to hold things together.

The nature of this force was (and to some extent still is) mysterious. In the first place, the magnitude of the force is totally unprecedented in nature. Although electrical repulsion between two positive charges is a well-known phenomenon, the fact that within the nucleus the charges are separated by only 10^{-13} centimeter or so introduces an entirely new scale into the discussion. We can get some idea of what this means by performing a little thought experiment. First, we calculate the force between two protons in a nucleus, and then scale everything up so that the nuclei are spheres a foot across (with their centers perhaps 18 inches apart). If the repulsive force were scaled up by an equivalent amount, how large would it be?

One way of gauging its magnitude would be to imagine imbedding the two scaled-up protons in a block of solid steel. Even if we made the block of the strongest alloy known, the electrical repulsion between the protons would rip them apart, tearing the steel block as if it were tissue paper. Whatever it is that holds the nucleus together would have to be many orders of magnitude stronger than steel or the nucleus could not exist.

The simple fact that the nucleus does exist leads us to the conclusion that there must be some process in nature that is capable of overcoming this repulsion between protons. The process must produce much stronger forces than anything we know about in the ordinary macroscopic world. Physicists have named this process the *strong interaction,* and refer to the force generated by this interaction as the *strong force.* The development of elementary particle physics has been an attempt to understand what the strong force is and how it is generated. For now, we just note that the existence of stable nuclei poses a major problem in our understanding of the atom.

What Is It Made Of?

THE next question that has to be asked about the nucleus, once we accept its existence, concerns its composition. We know that there must be protons in it, of course, because it has a positive electrical charge. But when you start to look at the different chemical elements and compare the mass of the atom with the number of protons the atom must have, an unexpected fact becomes evident.

The simplest atom—hydrogen—is no mystery. The nucleus of this atom is a single proton, and a single electron orbits around it. The single positive charge of the proton cancels the single negative charge of the electron so that the electrical charge of the entire atom is zero, as it is known to be. The next most complicated atom—helium—has two electrons circling its nucleus. If helium is to be electrically neutral, then it must have a nucleus that contains two positive charges. Thus, we would expect the helium nucleus to contain two protons and, since virtually all of the mass of an atom is in the nucleus, we would expect the helium atom to weigh twice as much as hydrogen. In point of fact it does not. The helium atom weighs *four* times as much as hydrogen, so the helium nucleus must weigh roughly as much as four protons, but have the electrical charge of only two.

It turns out that this situation is pretty much the same for all of the chemical elements beyond hydrogen. The mass of the nucleus is close to a multiple of the proton mass and can be written approximately as

$$M_{\text{nuc}} \cong A \times m_p$$

where A is called the mass number of the atom. For hydrogen, $A = 1$, for helium $A = 4$, and so on. The charge of the nucleus, however, is expressed by a different integer, so that if we denote the charge on the nucleus by Q and the charge on the proton by q, we get $Q = Z \times q$, where Z is called the atomic number of the atom. Z is equal to the number of positive charge units found in the nucleus and therefore it is also equal to the number of electrons found in orbits around that nucleus. The experimental fact is that for elements beyond hydrogen, $A \gtrsim 2Z$; that is, the mass number is always at least as much as twice the atomic number. In heavy elements, such as uranium, it can be as much as 2.6 times greater than Z. Regardless of the details of how A and Z behave in different chemical elements, the central message is clear. There is twice as much matter in the nucleus as is needed to explain the charge. What form does this extra matter take?

One of several hypotheses that were put forward was that there ought to be another, hitherto unknown particle that had about the same mass as the proton but which carried no electrical charge. This particle was called the *neutron*. Finding and identifying the neutron then became a problem for experimental physicists.

In 1932 the British physicist James Chadwick, working at Cambridge, England, was studying something called the radiation of beryllium. It had been discovered that when particles from naturally radioactive sources were allowed to strike a target made of a thin sheet of beryllium metal, a type of radiation came out of the metal. This radiation had no electrical charge. Using a device called a cloud chamber (which we will describe later), Chadwick saw that when the radiation of beryllium hit an atom of a gas such as nitrogen or helium, that atom would recoil, losing an electron in the process. Only particles with electrical charge can be detected in the cloud chamber, so what Chadwick saw looked like that depicted in Illustration 7. The radiation of beryllium would enter the device from the left, but since it had no charge it could not be seen. It would strike an atom, causing the recoil and loss of an electron. The atom would then have a charge, and its track would appear as a dark line.

Chadwick performed this experiment on atoms of different types, and for each type he measured the maximum recoil velocity of the atom. For example, he had data on collisions with helium, nitrogen, and oxygen atoms, among others. From these data and the known atomic sizes of the recoiling atoms, some simple laws of physics could be used to deduce the mass of the radiation of beryllium. When he carried out this calculation, Chadwick found that the unknown radiation consisted of particles whose mass was essentially the same as that of the proton. (Later measurements would show that it was a few percent heavier.) Thus, the radiation of beryllium could only be the missing piece of the nucleus—the neutron. For this discovery, Chadwick was awarded the Nobel Prize in 1935.

With this missing piece discovered, the atomic picture of matter seemed to satisfy the criteria, discussed in Chapter I, necessary for a truly simple picture. Every kind of material is composed of atoms, and

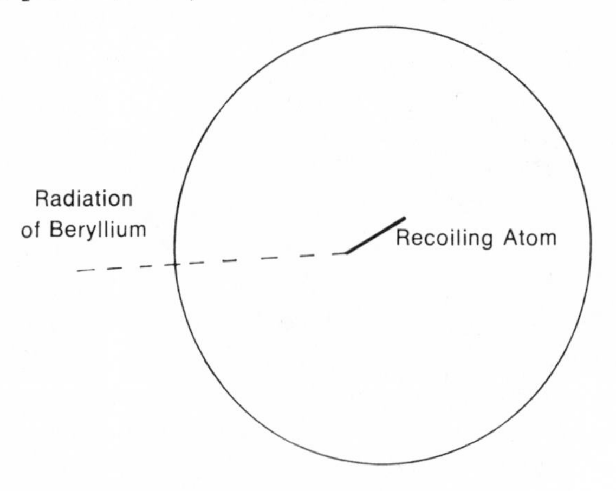

7. What Chadwick saw in the cloud chamber.

although there are many different sorts of atoms, they are all built out of three basic constituents, the proton, the neutron, and the electron. Of these three, only two, the proton and the neutron, are to be found in the nucleus. Because of this fact, they are often referred to collectively as *nucleons.* Of course, the question of how the nucleons could be held together inside the nucleus remained unsolved, but this was a relatively minor point compared to the enormous simplification that resulted from the development of the nuclear picture of the atom.

How the attempt to solve the puzzle of the strong interactions ultimately led to a breakdown of this simplicity is a story that will be told in later chapters. For now, there are some very interesting properties of the neutron that we should discuss.

The Instability of the Neutron

As mentioned in the previous section, the mass of the neutron turns out to be somewhat larger than that of the proton. In point of fact, the mass of the neutron is greater than the combined masses of the proton and electron. This larger mass means that it is possible (at least in principle) to make a proton and an electron from the amount of matter in a neutron. This, in turn, leads to one of the most striking properties of the neutron, its instability.

If (figuratively speaking) one were to take a neutron and set it on the table, the neutron would not remain there very long. In some hundreds of seconds, it would disappear, and in its place would be a proton, an electron, and another type of particle we will discuss later. In the language of particle physics, a neutron outside of a nucleus "decays," and the products of this decay include a proton and an electron. Since the neutron is the first example of an unstable particle we have encountered, we should take a little time to discuss this process of decay. Later on, we shall see that virtually all elementary particles share the characteristic of instability.

If we watched a large number of free neutrons, we would notice that they do not all decay at the same time. Rather, we would see one decay, then another, and so on at irregular intervals until all had completed the transformation into the decay products. It would be something like watching popcorn being made. Not all of the kernels "pop" at the same time: Each one goes when it is ready.

This means that if we started with 1,000 neutrons and made a graph showing the number of neutrons left at any time from the start, we would get a result something like the curve pictured in Illustration 8. The number of neutrons would decline steadily until, a long time from the start, none would be left.

In such a situation, we cannot say when an individual neutron will decay. There is no way of telling whether a particular neutron in our sample will be one of the first to go or the last. We can, however, talk about an average decay time for all of the neutrons in the sample by looking at the time it takes for a fixed fraction of the original neutrons to decay. One common way of describing decay times is to talk about the *half-life* of the sample. This is the time it takes for one-half of the original number of particles to decay. From the graph, we see that if we started with 1,000 neutrons at time zero we would have 500 after 636 seconds had elapsed. The half-life of a free neutron would therefore be 636 seconds.

Although the term half-life is associated with nuclear physics, the idea behind it is a rather common one in the sciences in general. It will arise, in fact, whenever the amount being removed from a sample depends on how much of the sample there is. To take one example, if there are two chemicals undergoing reactions in a solution, the amount of each chemical, if plotted as a function of time, will show a curve exactly like that shown in Illustration 8 for the neutrons.

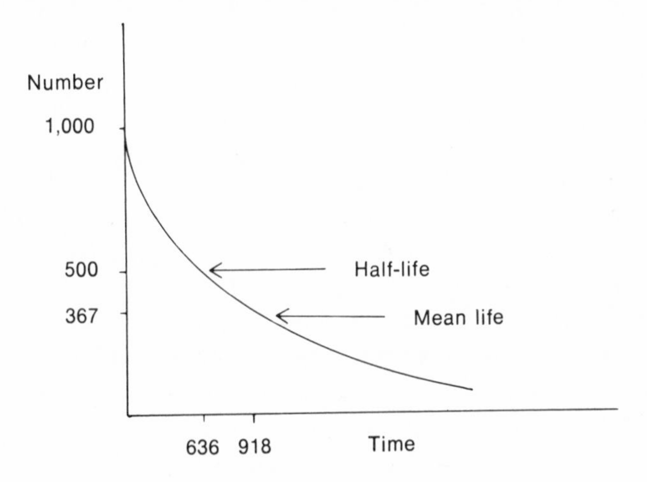

8. The number of neutrons left from a sample of 1,000 as a function of time.

Another common measure of the decay time of a particle is derived by looking at the time it takes for the sample to be reduced to $1/2.718 = 0.367 = 36.7$ percent of its original size. The number 2.718 is usually denoted by the letter e, and is known as the base of natural logarithms. It happens to be a number that arises naturally in the calculus and that appears regularly in laws describing natural phenomena. The time for the sample to become $1/e$ of its original size is called the *lifetime* of the sample, and if we denote it by the letter T, then the number of neutrons left at a time t after we start the counting will be

$$N = N_0 e^{-t/T}$$

where N_0 is the number of particles in the original sample. The lifetime of the neutron is 918 seconds (about 15 minutes).

If a neutron sitting by itself decays in a matter of minutes, how can the atomic nucleus exist indefinitely? Why, for example, do all of the six neutrons in the carbon atom not decay?

The answer to this question lies in a rather subtle point about nuclear structure. When a free neutron decays, the proton that is a result of that decay can go just about anywhere it wants, since there are no other protons around. In a nucleus this is not true. In order for one of the neutrons in a carbon atom to decay, there has to be room in the nucleus for the proton that would result from the decay. In carbon, and in most other nuclei, however, all of the places that the decay proton could fit into are already filled by other protons. (The details of how this works depends on something called the Pauli exclusion principle, which we will discuss in detail in Chapter XII.) The net result of this fact is that neutrons in most nuclei cannot decay, simply because there is no place for the decay products to go. To use an everyday analogy, the proton "parking lot" in the nucleus is filled, so the neutron cannot "park" its decay proton. Thus, most common nuclei are stable.

There are, however, nuclei where the neutron can and does decay. Such nuclei are unstable. When this happens, the nucleus emits an electron and acquires a unit of positive charge. The emitted electrons from such nuclei were one of the original types of radiation detected. They were called beta rays (it took some time to identify them as electrons); hence, this sort of reaction is called beta decay. The electron emitted in such a process has a great deal of energy and therefore travels a long way from its parent atom.

After beta decay we are left with an atom that has one more proton in its nucleus than it did before the decay. This means that the new atom has a net positive charge (since what happens in the nucleus does not immediately affect the electrons in their orbits). Usually such an atom will pick up a stray electron from the surroundings, becoming electrically neutral in the process. The net result of all this is a new atom, one which has one more proton and one more electron than its predecessor. Thus, the beta decay of a nucleus is one way in which the alchemist's dream of transmutation of the elements can be realized.

The Neutrino and the Weak Interaction

THERE is one aspect of neutron decay that needs closer attention. When the decay occurs, two charged particles are created—a proton and an electron. Thus, the total electrical charge of the final decay products is zero, which is exactly the charge on the neutron. This means that in the beta decay of the neutron the electrical charge is the same before and after the decay. We say that the electrical charge is *conserved* in the reaction.

This is an important point, because there are quantities that are not conserved. The number of particles, for example, changes from one (the neutron) to more than one in the reaction. Particle number is a quantity that is not conserved. Neutron decay is just one example of a general law that holds in all known elementary particle interactions.

In every reaction involving elementary particles, the total electrical charge is the same before and after the reaction. This is called the Law of the Conservation of Electrical Charge.

Conservation laws play an extremely important role in physics, so it was natural to ask whether other well-known conservation laws hold in beta decay. For example, there are laws that tell us that the energy of a system has to be the same before and after every reaction,* and other laws that tell us the same thing about momentum.

If E_0 is the total energy of the neutron before it decays, and if the beta decay is really described by the reaction $n \rightarrow p + e$, then it follows that if energy is conserved, the final energies of the proton and electron

*The reader who is unfamiliar with the concept of energy-mass equivalence will find it discussed in Chapter IV.

must add up to E_0. From this, it follows that if we look at two decays in which the protons have the same energy, the energies of the two electrons will also have to be equal.

It turns out that they are not!

If we looked at a large number of decays in which the protons all have the same energy, which we can call E_p, and plotted the number of times we saw an electron of a given energy as a function of energy, we would get something like the graph shown in Illustration 9. The electron energies would have all values from some minimum value E_{min} (which would correspond to the electron hardly moving at all) to the maximum value allowed by energy conservation, $E_0 - E_p$. The number of electrons at this latter value—which is what the law of conservation of energy says all of the electron energy ought to be—is actually zero.

What to do? There are only two ways of approaching a problem like this. One is to give up the law of conservation of energy—something physicists were very reluctant to do because the law seems to apply everywhere else in nature. The other alternative is to assume that there is another particle involved in the interaction—a particle that for some reason is not detected but which carries away the missing energy. In 1934 the Italian physicist Enrico Fermi (the man who later built the first nuclear reactor at the University of Chicago) put together the first successful theory of beta decay along these lines. Following some previous theoretical speculations, he supposed that beta decay is actually

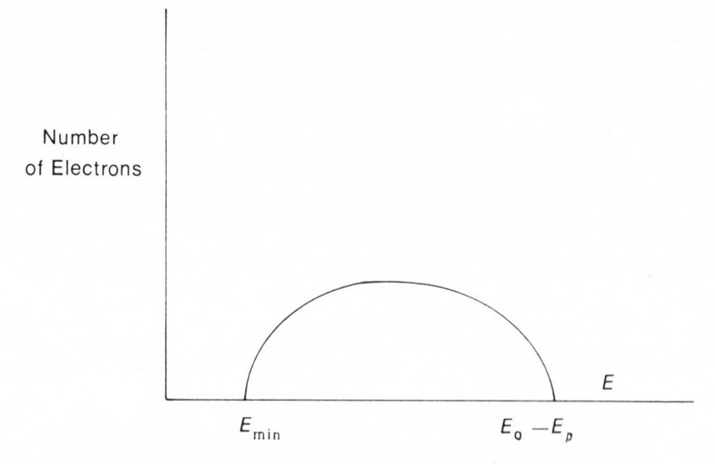

9. The number of electrons at each possible energy in beta decay.

described by the reaction $n \rightarrow p + e + \nu$, with the Greek letter ν (nu) standing for the new hypothetical particle. This particle must be electrically neutral (otherwise, it would have been detected earlier); hence, it was given the name *neutrino* (little neutral one). If such a particle existed, then it could carry away the missing energy from a beta decay and all of the accepted conservation laws of physics would be maintained intact.

Searching for the neutrino to verify its existence, however, turns out to be difficult. Because it is not electrically charged, it does not affect the electrons of atoms as charged particles do, so detectors such as the Geiger counter will not show its presence. Unlike the neutron, it is unlikely to cause visible recoils when it comes near a nucleus. For one thing, like the photon, it is without mass. More important, it simply does not interact very readily with other matter. One way of gauging this lack of interaction is to ask how thick a lead plate would have to be in order for a neutrino to stand a good chance of interacting with at least one lead atom. For the neutrino that results from Fermi's calculations, the answer comes out not in meters, not even in kilometers, but in *light-years!* In other words, if a tube of solid lead stretched from Earth to the nearest star and we started a neutrino down that tube today, it would emerge at Alpha Centauri more than 4 years from now without having disturbed a single atom in the tube.

This property of the neutrino posed certain difficulties for physicists who wanted to detect its presence. As we have seen, the only way that any particle can be detected is for it to interact with something and cause a change. For a charged particle, the change may involve creating some free electrons. For a neutron, it may involve creating a recoiling heavy ion. But no matter what the change is, there must be some sort of interaction taking place if the particle is to be observed.

If the neutrino is really as reluctant to interact with other matter as our examples indicate, then the only way it can be "seen" is for a large number of neutrinos to pass through a detector, so that the small probability of interaction is counteracted by the large number of neutrinos. This, in fact, is how the neutrino was first detected in 1956—almost 20 years after Fermi first worked out the theory of beta decay. The source of the "flood" of neutrinos was a nuclear reactor (the experimental details will be given later). The neutrino is now fully accepted as a particle, and, in fact, is routinely produced for experimental purposes in many large accelerator laboratories around the world.

Thus, the introduction of the neutrino saves some cherished principles of physics while enlarging the roster of elementary particles. Reactions such as beta decay, in which the neutrino participates, typically occur on a very slow time scale. They are, therefore, usually called *weak interactions* to distinguish them from the strong nuclear interactions, which presumably occur in much shorter times. The study of weak interactions has played an important role in the development of our knowledge of elementary particles.

A Remark About Stability

WITH the neutron we have entered a system where particles do not retain their identity, but decay into other particles after a time. *Every new particle we discuss from this point on will be unstable.* In fact, if we look at the massive particles, only the electron and the proton can exist by themselves for indefinite periods of time. Ultimately, therefore, every other particle must decay into some combination of photons, neutrinos, protons, and electrons.

Of course, like all statements of scientific fact, this statement is only as good as the observations that have been made to back it up. The statement that no one has yet seen a proton decay does not necessarily imply that no one ever will. All we can really talk about are those experiments that have been performed to search for the decaying proton and see what sorts of limits we can set on our knowledge.

One way of doing this is to ask how large the mean lifetime of the proton would have to be in order to be consistent with present experiments. Since there are a lot of protons around, a long mean lifetime (and the consequent small probability that a proton will decay while being watched) can be compensated for by watching a large number of protons. Right now, the accepted limit on the proton stability is this: The mean lifetime of the proton must be greater than 10^{30} years. From our discussion on neutrons, we know that this means that if we started out today with 1,000 protons, only 367 of these would be left 10^{30} years from now.

To get some idea of the time scale that would have to be involved in the decay of the proton, recall that the earth itself is about 5 billion years old (5×10^9), while the age of the universe is reckoned to be a bit more than 10^{10} years. This means that even if we started with 1,000 protons at the time of the formation of the universe 10^{10} years ago, not even one would have decayed by now! So when we say that the proton is stable what we mean is that its lifetime is at least many orders of magnitude greater than the lifetime of the universe, so that for all practical purposes the question of absolute stability is irrelevant.

By contrast, there are many naturally occurring radioactive substances with lifetimes of a billion years, the most familiar being uranium. Lifetimes of this length can easily be measured with modern techniques. Furthermore, unlike the proton and electron, decays of these atoms are seen in nature.

For reference, the lifetime of the electron is known to be greater than 10^{21} years, and, like the proton, no decay of the electron has ever been seen.

Provisional Box Score

WITH the discovery of the neutron, the Bohr-Rutherford picture of the atom seemed fairly complete. There were two elementary particles—the nucleons—which formed the nucleus of the atom. Around the nucleus a third elementary particle—the electron—orbited. These three particles taken together can be thought of as the fundamental building blocks of matter that the Greeks had hinted at and whose discovery we described as one of the basic goals of science. As we pointed out in Chapter I, a fourth particle—the photon—can be thought of as the constituent of radiation. As such, it is not actually a building block in the sense that a brick is a building block of a cathedral, but it is certainly a particle that is present in nature and that must be included in our list of what is elementary.

The unusual properties of the neutron—specifically its decay—led to the introduction (and eventual verification) of a fifth particle, the neutrino. Like the photon, it is without mass, and is not a building block in the structural sense of the word. Nevertheless, it is intimately tied both to the weak interaction and to the decay process of the neutron (which *is* a building block), so it must be added to the roster. The list of elementary particles is now up to five.

While the experimental investigations that led to the five particles were being carried out, the 1920s also saw the development of a theoretical framework to describe the physics of the atom. The new science was called quantum mechanics (in analogy to the study of the motion of macroscopic objects, which is called classical mechanics). The concepts that arise from this new theory are now pretty well understood and accepted in the scientific world, but they are sufficiently unusual for us to devote Chapter III to them.

CHAPTER III

A New Physics for a New World

No phenomenon is a true phenomenon until it is an observed phenomenon.

—JOHN A. WHEELER,
theoretical physicist

A man has to see, *and not just look.*

—LOUIS L'AMOUR,
The Quick and the Dead

The Physics of the Atom

SUPPOSE you were a Martian who liked to study languages and that after a long and difficult struggle you had managed to master German and French. Suppose that you had worked so long and so hard on these languages that you had come to believe that they were the only two languages that were spoken on Earth. Finally, suppose that because of your skill in language you were chosen to be a member of the first Martian landing expedition to Earth, and that your rocket ship put you down in the middle of North America. What would happen?

The first thing you would discover, of course, would be that the natives with whom you came into contact spoke a language that was very puzzling. Some of the words in the language (such as *hand*) would be exactly the same as German, and some of the other words (such as *cinema*) would be exactly the same as French. If you insisted that French and German were the only possible languages, you would have to conclude that the language you were hearing was a paradox, since it sometimes behaved like one and sometimes like the other. Philosophers back on Mars might even propound learnedly on the problem of "French–German duality" and ask whether the Martian mind was capable of understanding the true meaning of Earth's languages.

In this example, it is easy to see where the problem arises—our Martian friend has made the wrong starting assumption. There is no reason why a language has to be either German or French. Once he recognizes this fact, he will quickly see that English is a separate language with its own rules and vocabulary, but which happens to have similarities with both French and German.

In many ways, the map of the physical world is similar to a language map of the earth. Just as the languages spoken in different parts of the earth are different, so too are the laws that govern different aspects of physics. In Chapter II we saw how much smaller the atom and its constituents are than anything we encounter in everyday life. It should therefore come as no surprise that the behavior of subatomic particles is different from the behavior of particles in the larger world. In fact, assuming that circumstances in the atom have to be as they are in large objects with which we are familiar is something like the mistake our Martian linguist made when he assumed that every language spoken on Earth had to be one that he knew.

When physicists in the 1920s began to make a serious study of elementary particles, they were familiar with two sorts of things from their work with large-scale objects. They called these things by the names *particles* and *waves.* Particles are similar to baseballs; they are located at a specific region in space, they can move from one region to another, and they can be described in terms of their position and their velocity at any moment. Waves are also familiar to us (think of surf coming into a beach). They move from place to place, like particles, but they are not located at a specific point. The crest of a wave may move along with a given velocity, but the wave itself is spread over the extended region between successive crests. To describe a wave we therefore have to specify its velocity and the distance between crests, or wavelength. As you might expect from the differences in their natures, particles and waves in the large-scale world behave differently from each other.

With this sort of background, it is not surprising that when physicists began to investigate elementary particles they did not ask "What are they?" but "Are they particles or waves?" In this sense, they were like the Martian linguist who assumed that everything he encountered had to be like the things he already knew. And just as the Martian soon found something that did not fit into his conceptual scheme, so too did the first particle physicists. They found that electrons, which normally exhibit the kind of localization associated with particles, could also behave like waves under certain conditions. Similarly, light, which normally displays the behavior of waves, would start to look like particles. In the end, nothing in the subatomic world looked exactly like a particle or exactly like a wave, and this state of things was profoundly disturbing to the physicists. They coined the term *wave-particle duality* to express this feature of the objects they were studying, and philosophers picked up this term and used it to "prove" that there are inherent limits to what the scientific method can uncover.

But in terms of our Martian analogy, we recognize that wave-particle duality does not arise because of anything paradoxical about the behavior of elementary particles, but simply from the fact that we have asked the wrong question. If we had asked "How does an elementary particle behave?" instead of asking "Does it behave like a particle or a wave?", we would have been able to give a perfectly sensible answer. An elementary particle is not a particle in the sense that a bullet is, and it is not a wave like the surf. It exhibits some properties that we normally associate with each of these kinds of things, but it is an entirely new kind of phenomenon. In this sense, it is like English in our Martian analog—it has aspects in common with familiar things, but it is neither of these things. It is something new.

Of course, it is easy for us, looking back to the 1920s, to see how pointless all the debate about the wave-particle duality was. We should also realize that taking this attitude is a sort of Monday morning quarterbacking that detracts nothing from the achievement of the men who first unraveled the laws of the subatomic world. And lest we start feeling too cocky, remember that 50 years from now someone may be pointing out how pointless some of *our* great debates are.

Once we have understood that the electron is not a particle and not a wave, we should ask "What is it?" A full answer to this question would take a long discussion of quantum mechanics, which we lack the space to present here, so I shall summarize some of the results. When physicists say that something is a wave, what they really mean is that the thing's behavior can be predicted according to a particular equation called, appropriately enough, the wave equation. For a wave on water, for example, this equation will predict all of the relevant properties of the wave—how fast it will move, how high above the normal surface the water will be at any particular time, and the shape of the wave. These are illustrated in Illustration 10. As the wave moves by a point in the water, the water rises, finally reaching the full height of the wave. This quantity is labeled *A* in the illustration, and is called the amplitude of the wave. It is one number that characterizes the wave.

Another quantity that characterizes the wave is its shape. One way of indicating the shape of the wave is to say how high the water is at a given time for each point along the wave. One such height is labeled *D* in the illustration. The height of the wave at an arbitrary point is called the displacement, and it follows that the maximum value of the displacement is the amplitude of the wave.

The wave equation is a relation written in the language of differential calculus. It tells us the connection between the displacement of the wave, the time, and the properties of the medium on which the wave moves. Since there is no reason why a wave on water should have the same properties as a similarly shaped wave in alcohol, it is the properties of the medium that ultimately determine the speed of the wave. But the end result of the mathematical analysis of the waves is that if you give a physicist the shape of a wave at one point in time, he can then predict where that wave will be (and what its shape will be) at any future time.

It was therefore only natural that when evidence started to accumu-

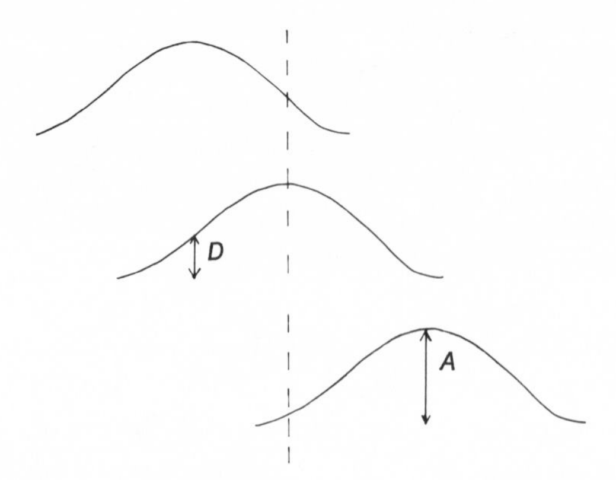

10. The important properties of a wave.

late, which indicated that the electron had some properties normally associated with waves, someone would try to describe the electron by the wave equation. In 1926 Erwin Schrödinger, an Austrian theoretical physicist, wrote down such an equation, calling the displacement of the electron "wave" the *wave function* and denoting it by the Greek letter ψ (psi). Even though he did not know what the wave function signified in terms of the properties of the electron, Schrödinger found that he was able to use his equation to solve many of the outstanding scientific problems of the time—the structure of the hydrogen atom, for example. Thus, the statement that the electron exhibits properties normally associated with waves was transformed by Schrödinger into a precise mathematical description of the electron "wave."

But what is the wave function? Some physicists at the time wanted to interpret the displacement of the electron wave as being real in the sense that the "true" electron was spread out and the displacement measured how much of the electron was at each point in space. It remained for Niels Bohr in Copenhagen to provide what is now the accepted interpretation of the wave function. He reasoned that there was too much evidence for the particlelike properties of the electron to allow it to be smeared out in a classical wave. The electron, he said, should still be thought of as a localized object, but the displacement of Schrödinger's electron wave at a particular point is related mathematically to the *probability* that a measurement would show the electron to be located at that point. In this interpretation the Schrödinger equation predicts the properties of a *probability wave,* and with it we can predict the probability that an electron will be at a certain point if we know the wave function.

One possible picture of what an elementary-particle wave function might look like is given in Illustration 11, below. As this wave moves past a particular point, the probability of finding the electron at that point will change. We can calculate how much it will be at any instant by using Schrödinger's equation to find the wave function, and then getting the probability from the relation $P = \psi^2$.

A wave that is bunched up and confined to a specific region of space, such as the one pictured in the illustration, is called a wave packet. The description of most elementary particles is in terms of this sort of wave. In this picture, it is clear that the particle has both particle and wavelike properties, since the wave packet is confined to a relatively small region

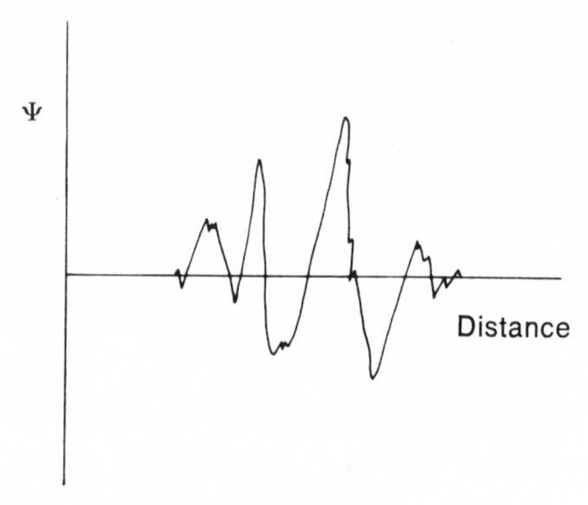

11. Possible picture of an elementary-particle wave function.

of space (a property normally associated with particles), but the packet itself is a wave. In addition, the picture allows for the fact that an actual measurement will show a pointlike particle somewhere in the region of the packet.

But isn't this whole interpretation just saying that the electron is really a pointlike classical particle, and isn't the probability interpretation just a statement about the fact that we simply do not have good enough instruments to detect the electron's actual location? It is tempting to think that such an objection might restore a familiar view of the subatomic world, but there are some problems with making measurements on the subatomic level that lead to even stranger results than the idea of a probability wave. It is to these problems that we now turn.

The Uncertainty Principle

PHYSICS is a science which, in the end, rests on our ability to observe and to measure the world around us. The plain old nuts and bolts of measuring a quantity accurately have often made crucial differences in the development of science. For example, in 1572 the Danish astronomer Tycho Brahe, using newly designed instruments that were more accurate than any that had been used up to that time, showed that a nova seen in the sky was actually as far away as the other stars, thereby exploding the Aristotelean idea that the heavens, being perfect, could not change. The concept of measurement also occupied a large place in the thinking of the men who developed the quantum mechanics.

We must make one distinction very clear—the distinction between the measurement itself and the accuracy with which the measurement is made. For example, suppose you were going to measure your height. You could do it by standing up against a wall, having someone mark off your height on the wall, and then measuring the height of the mark on the wall. If you carried out this procedure, you would get a number—6 feet, for example—that would represent the result of the measurement.

How accurate would this result be? One way of finding out would be to do the measurement again. If you did this, you might be slouching a little more the second time, or your helper might make a slightly different estimate of the spot on the wall opposite the top of your head. There are many things in the measuring procedure that might be different the second time around, so you would expect the result you get to be a little different, too. You might, for example, find your height to be 6 feet 1/8 inch. A third try might give 5 feet 11¾ inches, and a fourth try 6 feet again. If you made a graph by plotting the number of times you measured a certain height against the height measured, you would get something like what is pictured in the left portion of Illustration 12.

A bar graph of this type is called a histogram, and this particular histogram shows two things. First, it shows that the measurements all cluster around 6 feet, and, second, that 6 feet is the average of all of the

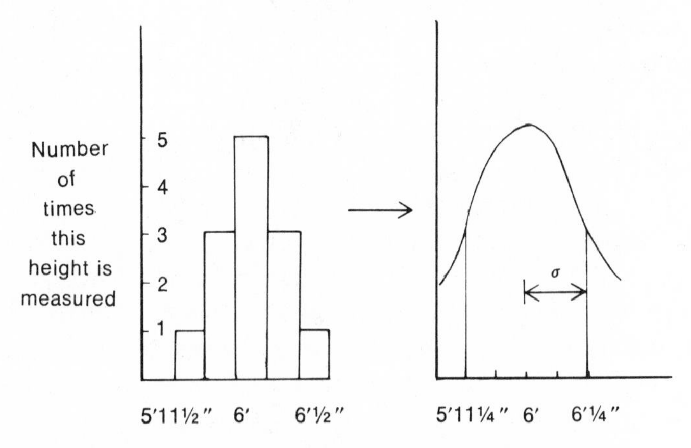

12. Results of measurement of height discussed in the text. The histogram on the left turns into the curve on the right when many readings are taken.

measurements. Thus, you would report that your height was 6 feet even.

But the histogram also shows that there is a scatter in the measurements; that they vary from one trial to the next. The spread in these readings gives us some idea of how accurately the measurements were made. In the histogram in the illustration, for example, there are values of the height obtained that are 1/2 inch above and 1/2 inch below the average, with most readings being within 1/4 inch of the average. From this you can conclude that your height is not 5 feet 11 inches or 6 feet 1 inch, that it probably is not 5 feet $11\frac{1}{2}$ inches or 6 feet 1/2 inch, but that it might be 5 feet $11\frac{7}{8}$ inches or 6 feet 1/8 inch. The spread in the measurements shown on the histogram indicates something about the certainty with which we can claim our average measurement is correct or, equivalently, what the uncertainty of the measurement is.

We can give a more precise definition of uncertainty by imagining our taking a very large number of height measurements. In that case, the histogram would develop into the continuous curve shown on the right in Illustration 12. This curve is called a normal or Gaussian distribution of measurement results. The quantity we have labeled by the Greek letter σ (sigma) on the graph is called the standard deviation. It is defined as that distance from the center of the curve within which 68 percent of the measurements fall. Hence, if the graph represents 1,000 height measurements, then 680 measurements will be between $+\sigma$ and $-\sigma$. In this example, $\sigma \cong 1/4$ inch.

The standard deviation of a large number of readings can therefore be used as an indication of the accuracy of the experiment and, by the same token, as a measure of the uncertainty of our knowledge of the true value of the quantity being measured. In the example of measuring height that we have been using, the existence of uncertainty is due to a number of factors that can vary from one measurement to the next —your posture, the judgment of your helper, how well you can read a ruler, and so forth. There would be little difficulty, however, in designing the experiment so that the uncertainty would be smaller than 1/4 inch. You could, for example, take all measurement while lying down to eliminate the posture problem, and use accurate levels to locate the top of your head. There is no reason, in principle, why you could not measure your height to 1/100 of an inch or even 1/1000 of an inch. In fact, in the ordinary world of classical physics, there is no reason why the uncertainty of this measurement could not be zero. The only limitation has to do with how expensive and accurate your instruments are.

The second important point about the measurement of your height is that you will be the same person after the measurement as you were before. What I mean is this: If you wanted to know your weight in addition to your height, you could, in principle, know it with zero uncertainty as well. Furthermore, the fact that you had measured your height would not affect the measurement of weight. In the language we have been developing, in the classical world it is possible to measure both height and weight with zero uncertainty.

In the world of the atom the situation is different in one important respect. While it is still possible, in principle, to measure a quantity with zero uncertainty, the process of carrying out the measurement may make it impossible even in principle to measure another quantity to the same degree of accuracy. It is as if measuring your height to high accuracy meant that you could measure your weight only to a very low accuracy or vice versa. Let us see how this rather surprising state of affairs comes about.

When we talk about observing or measuring something in the macroscopic world, we always have in mind (at least implicitly) some method by which we can cause something to interact with the measured object. In the height measurement, for example, your helper either looked at the top of your head to see where to mark the wall, in which case he was seeing light that had interacted with you, or he laid a marker directly on your head. The entire discussion of height rested implicitly on the assumption that the interaction with the measured object did not change it in any important way—that the act of observation itself could be neglected. And, of course, when we talk about macroscopic objects, this is a perfectly reasonable assumption. No one would seriously suppose that looking at someone's head changes his height. The idea is too ridiculous to take seriously.

Why?

If pressed for an answer, you would probably say something like this: The photons that make up the light are so small and have so little energy that there is no possible way in which they could affect a large object, such as a human being. This is precisely the way that classical physicists have always justified the assumption that the observer does not affect the observed, and the assumption is reasonable when what is being observed is huge compared to what is interacting with it.

This argument will not work for elementary particles! In order to "look" at an electron, we must cause it to interact with a photon or some other particle. In either case, we can no longer argue that the interaction can be ignored, since the object being measured and the probe we use upon it are both of the same size and energy. This simple fact is the physical basis for one of the most celebrated laws of quantum mechanics, the uncertainty principle. Discovered in 1927 by Werner Heisenberg, this law states that there are situations in the subatomic world where it is not possible, even in principle, to know the values of two different quantities relating to an elementary particle because the act of measuring the first interferes with our ability to measure the second. Before giving a precise definition of the uncertainty principle, we can illustrate how it might work in our height-weight analogy.

Suppose we wanted to know someone's height and weight, but for some reason the only way we could measure height was by the following bizarre technique: The person to be measured would lie down, and then someone would drop small weights near him from above. When the weights fell straight down past the person, we would know we were in the region above his head, and when they were deflected we would know that we were below his head. By looking at the boundary between these two situations, we would have a measure of his height. (Although this probably would not be a good way to measure height, it is, in fact, quite close to the way in which the sizes of atoms and nuclei are measured.)

In the process of dropping weights, some weights would stay on the subject and some would slide off; therefore, when we finished we would not know how many weights had stuck. If we now weighed the subject, we could surely measure a weight, but this would be the man's weight plus the weight added by the measurement of height. In this example we would therefore say that measuring the man's height caused an uncertainty in the measurement of his weight, and that the processes of height and weight measurement interfered with each other. I leave it to you to think of a way of measuring weight that would interfere with subsequent measurements of height.

The uncertainty principle is stated as follows: If we call Δx the uncertainty in the position of an object and Δp the uncertainty in its momentum (for our purposes, momentum is simply the mass of an object multiplied by its velocity), then in any attempt to measure these two quantities, the product of the uncertainties is given by $\Delta x \,.\, \Delta p > h$, where the symbol $>$ means greater than, and h is Planck's constant. In units where mass is measured in grams and length in centimeters, h has the value $h = 6.62 \times 10^{-27}$.

There are a number of things to note about this principle. In the first place, Planck's constant is a very small number; thus, when we are talking about macroscopic objects, the limits that the uncertainty principle places on our measurements are very small indeed. For example, if we have a 300-gram object, such as an apple, and we can determine its position to within one millionth (10^{-6}) centimeter, then the uncertainty principle states that the error in determining the velocity must be greater than

$$\Delta p > \frac{h}{10^{-6}} = 6.62 \times 10^{-21}$$

$$\Delta V > \frac{6.62 \times 10^{-21}}{300} = 2.2 \times 10^{-23} \text{ cm/sec}$$

Saying that the uncertainty in the measurement of velocity must be larger than this number makes very little difference in the macroscopic world, since any real measurement would produce uncertainty of the ordinary type which would certainly be many orders of magnitude larger. Hence, the smallness of Planck's constant provides a kind of justification for ignoring the interaction of observer and observed in the macroscopic world.

If the object we were considering were a proton, however, and the uncertainty in position were 10^{-8} centimeter (you will recall that this is about the size of the atom), then the uncertainty in velocity would be

$$\Delta V > \frac{h}{m \Delta x} = \frac{6.62 \times 10^{-27}}{1.7 \times 10^{-32}} = 3.9 \times 10^{5} \text{ cm/sec}$$

where we have used the result, from Chapter II, that the proton's mass is 1.7×10^{-24} gram. In this case, the uncertainty in the velocity is quite large, probably as large as the proton's velocity itself. Thus, on the quantum level, the uncertainty principle begins to be quite important.

Perhaps the best way to discuss what the uncertainty principle implies is to list some of the things that it does not imply.

1. *It does not imply that the particle's position cannot be measured exactly.*

You often run across the statement that the uncertainty principle proves that it is impossible to measure the position or momentum of a particle exactly, and from this statement innumerable incorrect conclusions are drawn. A quick look at the equation that defines the uncertainty principle shows that this is simply not the case. Measuring the position of a particle exactly would imply that the uncertainty in the position, Δx, would be zero. There is nothing in the uncertainty principle that says that Δx cannot be zero. All it says is that *if* $\Delta x = 0$, then Δp, the uncertainty in the momentum, must be infinite. If this were the case, then it would still be possible for the product of Δx and Δp to be greater than h. What this possibility implies is that if we decide that we want to know the position of a particle exactly, the measurement that we have to make to get this information will so alter the particle that we will be unable to get any information at all about its momentum. The resulting lack of a precise answer means that the momentum could be any number at all, which is just another way of saying that the uncertainty in the momentum is infinite.

2. *It does not imply that the particle "really" has a position and momentum, but that we just cannot measure them both.*

The uncertainty principle does such violence to our common sense feelings about what the world should be like that there is a strong temptation to think of it as a kind of technical limitation on what we can measure, rather than as a statement about a fundamental difference between the macroscopic and microscopic worlds. One therefore wants to think of the particle as if it "really" had a position and momentum (just as a billiard ball does), and to dismiss the uncertainty principle as a sort of pettifogging technical detail.

However, you have to ask what meaning you can give to a statement such as "The particle really has a precise position and momentum at any time, but we just cannot measure them both." There is no experiment one can imagine that could test this statement, because, as we have seen, measuring one variable puts a limit on how accurately we can determine the other. I would suggest that making a statement that cannot be verified by experiment, even in principle, is not a particularly useful way to do science. Whether or not the particle really has precisely defined values of position and momentum can make no difference in anything we can know about the subatomic world, so this sort of conjecture is in a class of statements that are not even right or wrong—they are just meaningless.

3. *If we could find some probe whose energy was very small compared to elementary particles, then the uncertainty principle would go away.*

This idea is known as the hidden variable theory of quantum mechanics. From our discussion of the physical basis of the uncertainty principle it seems fairly clear that if a new regime of sub-subatomic particles were discovered which could interact with the electron (for example) in the same way that light interacts with macroscopic bodies, then measurements made with these new particles would not disturb the electron and we could get around the uncertainty principle.

This is true, but people who work out the consequences of hidden variable ideas have been able to prove that if such particles existed, there are certain experiments in which ordinary quantum mechanics and hidden variable theories would predict different results. The experiments that have been done tend to bear out the ordinary quantum mechanics; so, while it is always dangerous to say that something in physics is impossible, it seems to me that it is very unlikely that anything resembling a hidden variable theory will eventually undo the uncertainty principle.

Virtual Particles and the Strong Interaction

In the previous section we talked about the uncertainty principle in terms of two variables—position and momentum—because these are quantities that are familiar to us all. It turns out that the uncertainty principle applies to a number of other pairs of variables as well. In the development of the ideas of elementary particles, the most important of these variables are energy and time. If we denote by ΔE the uncertainty in our knowledge of the energy of a quantum system and by Δt our uncertainty about the time at which it has a given energy, then reasoning that is precisely like that leading to the position-momentum relation leads to the equation $\Delta E \,.\, \Delta t > h$, where, again, h is Planck's constant. All of the remarks about the implications of the uncertainty principle made in the last section apply to this relation as well.

To interpret this type of uncertainty relation, let us think for a minute about what is involved in measuring the energy of a system. As in any other measurement, we can only observe the system by using some sort of probe. In order to find the system's energy, this observation will take a certain amount of time (at the very least, it will involve the time it takes for the probe to interact with the system). We can interpret Δt, the time uncertainty, as being the time it takes to make the observation, since obviously we are totally unable to answer questions about the energy in any lesser time scale.

Suppose, for example, that we were measuring a system in which it took our instruments 1 minute to take a measurement of the energy. We could then report this energy as our result, but if someone asked us whether our reported energy was the actual energy of the system at 10 seconds or at 50 seconds, we could not say. All we could say would be that during that minute, the average energy was as we reported it. It is even possible that the actual energy of the system changed during the minute (see Illus. 13), so that sometimes it was higher than the average and sometimes lower. In this case, the fact that the time determination was uncertain would lead to uncertainties in the energy as well, since the energy could have been as high as E_{max} or as low as E_{min} during the interval. In classical macroscopic systems, of course, both the time and the energy uncertainties can be reduced simultaneously to zero (at least in principle).

In the quantum world, however, that reduction is not possible. If the

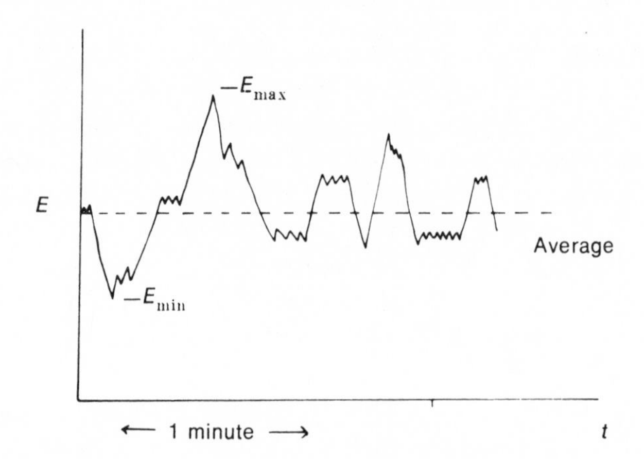

13. Possible energy of a system as a function of time.

probe's interaction is instantaneous (so that you know the time exactly and $\Delta t = 0$), the interaction is not long enough to give any indication of the energy, so $\Delta E = \infty$. If we observe for a long time in order to measure the energy exactly, then $\Delta t = \infty$ and $\Delta E = 0$. As with momentum and position, you can measure one or the other exactly, or both with some uncertainty, but you cannot get them both exactly.

The energy–time uncertainty relation leads to a very interesting concept in physics—the concept of the virtual particle. In Chapter IV we shall discuss the implications of the famous relation between energy and mass first written down by Einstein, $E = mc^2$. For our discussion here, we only have to understand that the energy associated with the mass of a particle can be included in the uncertainty relation as well as any other energy. In particular, if a particle of mass M is sitting by itself and we measure its energy in a time Δt, there is an uncertainty in the mass of the system of the amount

$$\Delta M > \frac{h}{c^2 \Delta t}$$

If Δt is small enough, it is even possible that the uncertainty in the mass may be large enough so that during the time Δt we cannot tell whether there is a single particle of mass M or a set of particles of total mass $M + \Delta M$ sitting at a particular point in space. In other words, we could have a sequence such as the one pictured in Illustration 14, below, where a single particle becomes, for an instant, a pair of particles. No measurement we can make will tell us that this is or is not happening. We say that the original particle "fluctuates" into two particles in this process, and we call the extra object a *virtual particle.*

To get some idea of the times we are talking about for this fluctuation process, we can ask how small Δt would have to be in order for a single proton to fluctuate into a proton and an additional particle that has the

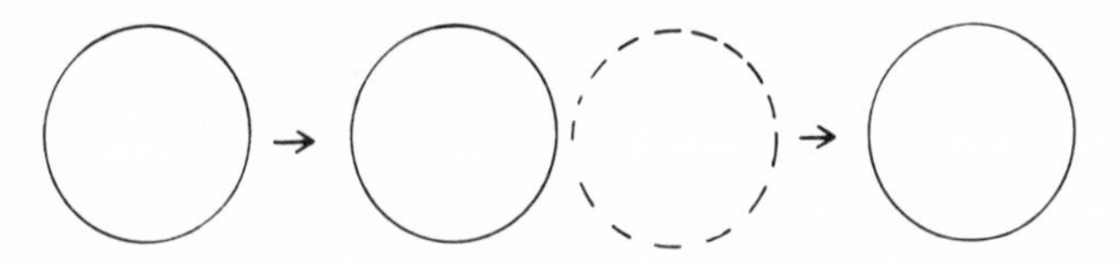

14. A particle *(left)* fluctuates to itself plus a virtual particle *(center)* and back to itself again.

same mass as the proton. For this sort of fluctuation, the mass uncertainty is

$$\Delta M = M_p = 1.7 \times 10^{-24} \text{ g}$$

and, recalling that the speed of light is $c = 3 \times 10^{10}$ centimeters/second, we find

$$\Delta t > \frac{6.6 \times 10^{-27}}{9 \times 10^{20} \times 1.7 \times 10^{-24}} = 4.3 \times 10^{-24} \text{ sec}$$

The speed at which the virtual proton can travel is limited by the velocity of light, and it might be interesting to know how far it could travel at that speed in the time Δt. We find

$$d = c\Delta t = 3 \times 10^{10} \times 4.3 \times 10^{-24} = 1.3 \times 10^{-13} \text{ cm}$$

—which is about the size of the proton itself.

This calculation has a very interesting interpretation. One way of visualizing the appearance of a virtual particle is to think of it as "sneaking out" while no one is looking. So long as it gets back "home" before a time Δt has elapsed, the uncertainty principle guarantees that no one will know the difference. In a sense, the principle plays the same role in this process as the clock played in the Cinderella story—so long as the particle gets home from the "ball" before the time runs out, it will not turn into the subatomic equivalent of a pumpkin.

The distance that the virtual particle can travel in its alloted time is also an interesting quantity. Suppose that we had two ordinary particles sitting a distance d_v apart. We could then have a situation in which one particle fluctuates into itself plus a virtual particle, and, provided that the mass of the virtual particle is such that it can travel a distance d_v during its life-span, the virtual particle could be absorbed by the second of the original two particles. (See Illus. 15.) This is called the exchange of a virtual particle, and, again, the uncertainty principle tells us that such processes can go on without our being able to detect any violation of the law of conservation of energy.

From our example, we know that a virtual particle whose mass is the same as that of the proton can travel about 10^{-14} centimeter, even if it moves at the speed of light. If, on the other hand, the virtual particle has a mass that is a fraction of that of the proton, it can travel several times this

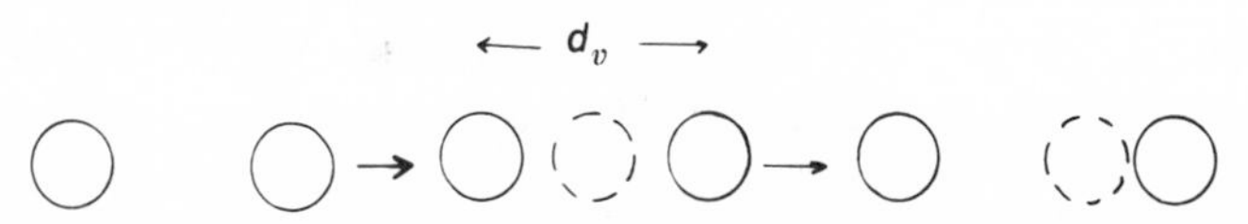

15. How two particles can exchange a virtual particle and still not violate energy conservation.

distance. This means that a virtual particle of such a mass can be exchanged between protons that are several fermis apart, and from Chapter II, we know that several fermis is about the size of an average nucleus.

In 1934, the Japanese physicist Hideki Yukawa, working at the Osaka Imperial University, wrote a landmark paper that showed that if two protons exchanged virtual particles, the result of the exchange would be an attractive force between the protons. Furthermore, he showed that if the virtual particles had a mass something like 1/9 of the proton mass (the exact fraction was unimportant), then the result would be a force strong enough to overcome the electromagnetic repulsion between the protons and would, in fact, tend to hold them together. In other words, the exchange of such a virtual particle could generate the strong force needed to hold the nucleus together.

To understand how an exchange of particles could result in a force, consider what would happen during the time the virtual particle is in transit if it exerted an attractive force on the two nucleons. During the brief moment of existence of the virtual particle, each nucleon would be pulled toward it. The net result would be that each nucleon would be attracted toward the other, prompting an outside observer who could not see the virtual particle exchange to say that there was an attractive force between the nucleons. In Yukawa's hypothesis, the new particle had precisely the properties needed to produce such an effect.

Of course, no particles with mass less than the proton and more than the electron were known at that time, but Yukawa suggested that it might be a good idea to look for what he called the *U* quantum. His hypothetical particle was later termed the *meson* (intermediate one) because of its mass. Although the strange story of the search for this particle will be told in a later chapter, let us spend a minute on the implications of this new idea of the strong force.

In the first place, the nucleus is now seen not as a static collection of protons and neutrons locked into place like some sort of Tinker Toy construction, but as a dynamic system in which mesons are whizzing around from particle to particle. They are continuously being created at one spot and absorbed at another, and this process is what holds the nucleus together. In a sense, the mesons form the "nuclear glue" that we saw was needed in order to understand nuclear structure.

Since both protons and neutrons can emit and absorb mesons, it is easy to see the role that the neutrons play in the nucleus. They have no charge, so they do not contribute to the electromagnetic forces that are pushing the nucleus apart; but they can contribute to the forces binding it together. Thus, they play the dual role of diluting the repulsive forces and enhancing the attractive ones. It should therefore come as no surprise to learn that most stable nuclei have as many, or more, neutrons as protons.

Perhaps more important than the explanation of the strong interaction given by Yukawa is the idea that we can think of a force as being due to the exchange of particles. Although we have discussed this idea only for the strong interaction, it should not be too hard to accept that modern physicists think of *all* forces as being due, ultimately, to the exchange of particles. The electrical force, for example, is thought to arise because of the exchange of virtual photons; the gravitational from the exchange of particles called gravitons (no gravitons have been detected as of this writing). So the principles of quantum mechanics as developed by Yukawa have led us to a whole new way of looking at forces in nature—a way that is intimately tied to the structure of matter and to elementary particles.

Finally, lest the reader leave this discussion with the impression that the whole thing is some sort of elaborate shell game, I should point out that modern nuclear physics experiments provide strong (albeit indirect) evidence for the existence of virtual particles. When high-energy electrons are allowed to collide with nuclei, the results of the collisions seem to indicate that although the electrons usually encounter nucleons, one of them occasionally happens to be in the right place at the right time to hit one of the virtual particles during its transit. In this sense, the virtual particles are really there.

CHAPTER IV

Energy, Matter, and Antimatter

Things equal to the same thing
are equal to each other.

—EUCLID,
Elements

Mass and Energy

ALTHOUGH the results of the theory of relativity are many, one of them has become so familiar that it could almost be classified as a piece of folklore: It is the famous relation between energy and mass, $E = mc^2$. We have already seen that this equation has far-reaching consequences in the world of elementary particles, leading ultimately to the idea of virtual particles and the concept of force as a result of the exchange of such objects. In this chapter we will see that this equation also leads to another consequence for elementary particles—the existence of antimatter. But before we go on to that subject, we should take a moment to discuss the most common question that people have about the Einstein equation: "Why is it the speed of light figures in this equation and not something else, such as the speed of sound?"

Actually, this question goes right to the heart of the theory of relativity, which is where the Einstein equation comes from. The basic postulate of the theory says that the laws of physics have to appear to be identical to every observer, whether he is moving or not. Thus, relativity places a heavy emphasis on the laws of physics, and anything that is built into these laws will occupy a special place in the theory.

The laws that govern the fields of electricity and magnetism are called Maxwell's equations, after James Clerk Maxwell, the British physicist who first wrote them down in 1873. It turns out that these equations predict that there will be waves that can move through the vacuum, and that the speed of these waves is related to measurable quantities, such as the force between electrical charges and the forces exerted by magnets on each other. What led Maxwell to identify these new waves with light was the fact that when he took the measured forces and put the appropriate numbers into his theoretical value of the wave velocity, it came out to be 3×10^{10} centimeters/second, or 186,000 miles/second, which is precisely the measured velocity of light. In this sense, the speed of light plays a special role in physics because it is built into the laws of electricity and magnetism, whereas no other speed plays a similar role in any other branch of physics. Basically, this is the fact that leads to the c^2 term in the mass-energy relationship.

I must make one other point before we leave this topic. It is customary to refer to c as the *speed of light,* which, of course, it is. A more correct terminology, however, would be *speed of electromagnetic radiation.* Visible light is only one type of an enormous variety of such waves, and all of them move with velocity c. This family includes infrared and ultraviolet light, radio waves, X rays, microwaves, and gamma rays. All of them can be thought of as consisting of photons, although the wavelength of the photon corresponding to each type is different. All move at the same speed, and visible light has no special place other than as an example of the entire class of electromagnetic radiation.

Energy is defined as the ability to do work and work is defined to be the product of a force times the distance through which that force acts. Thus, energy must ultimately be related to the ability to produce a force that can act through a distance. For example, if you lift a weight, your muscles supply a force that overcomes the force of gravity exerted on the object you are lifting and that acts through the distance through which the object is lifted. You have done work (force $\times$ distance) on the object, and consequently the object has thereby also acquired the ability to do work. For example, if the object fell, it could exert a force on an object in its path. Thus, when we lift an object, we do work on it, but once it is lifted it can do work on something else. It therefore has energy. This energy is associated with the object's position—the higher it is, the more energy it has; this is called *potential energy.*

In this example, we have talked about the energy an object has because it has been lifted in the presence of a gravitational field. There are other kinds of potential energy associated with other kinds of forces. For example, if the electron in the atom shown in Illustration 16 is originally at the point *A* and we cause it to be moved to the point labeled *B,* we have to do work to overcome the attractive electrical force between the nucleus and the electron. The electron has acquired potential energy in this process, just as the object acquired energy when it was lifted against the force of gravity. Consequently, moving electrons around near a nucleus, or rearranging electrons in a set of atoms, will change the energy of the system. Since this type of rearrangement takes place in chemical reactions, this type of energy is sometimes referred to as *chemical potential energy* to distinguish it from the gravitational variety. When you burn gasoline in your car, for example, the energy to run the car comes, ultimately, from the chemical potential energy liberated when long hydrocarbon molecules are broken up into smaller ones, with an accompanying rearrangement of electrons (see Illus. 16).

There can also be energy associated with motion, and this is called *kinetic energy.* For example, you have to apply a force with your hand over a distance to get a bowling ball rolling. When the ball hits the pins, some of them are knocked into the air (acquiring some potential energy in the process) and some just move away from the ball. Clearly, you imparted energy to the ball with your hand just as surely as you impart energy to something that you lift against the gravitational field.

A somewhat more subtle form of kinetic energy is evidenced by the

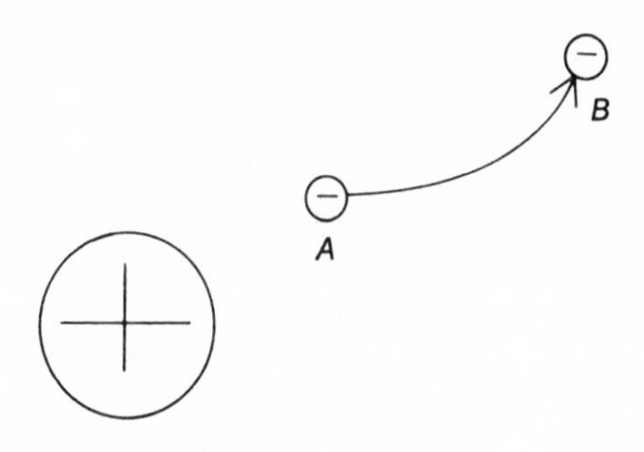

16. Work is done in moving the charge from *A* to *B*.

process of lighting a fire by rubbing two sticks together. On first glance, it would seem that the kinetic energy of the sticks was dissipated in the rubbing process, but if you could look closely at the atoms in the sticks, you would find that as the temperature of the wood increases, the atoms move faster and faster. Thus, heat energy can be thought of as energy of motion at the atomic level.

By the end of the nineteenth century, physicists knew of two general types of energy—kinetic and potential—and of many subclasses of each. In addition, they knew that, while it was often possible to change the form of the energy in a system, the energy in an isolated system had to remain constant in time. For example, we discussed how the chemical energy in gasoline can be converted to the kinetic energy of an automobile. This is a process in which energy changes form, but the energy of the car will always be less than (or at best, equal to) the change in potential energy in the gasoline. This conclusion, which tells us that energy can neither be created nor destroyed, is called the law of conservation of energy, and it is one of the foundation stones of classical physics.

The theory of relativity does not change this principle, but it does add a new category for energy. In addition to potential and kinetic energy, Einstein's equation tells us that mass is also a form of energy. And just as energy can be converted from potential to kinetic and back again, so too can energy be converted from familiar forms to matter and back again.

Does this mean that it is actually possible to create matter where none existed before? Einstein's equation tells us that the answer to this question must be yes! In fact, much of the rest of this book will be devoted to a study of what happened when this process of creating matter began being performed routinely in laboratories all over the world.

From the fact that the velocity of light is a very large number, it follows that it takes an enormous amount of energy to make even very modest amounts of mass. The unit of energy in the system where mass is measured in grams and length in centimeters is called the erg. A 100-watt light bulb uses about 1 billion ergs each second when it is operating, so the erg is not a large energy unit. Each day Americans use about 2.3×10^{24} ergs for all of their energy requirements. This includes home heating and lighting, transportation, electricity, and industry.

According to Einstein's formula, how much mass would we need to convert to energy in order to supply this energy? From the equation, the mass is

$$m = \frac{2.3 \times 10^{24}}{9 \times 10^{20}} = 2.5 \text{ kg} \approx 5 \text{ lb}$$

In other words, if we could convert matter to energy with 100 percent efficiency, 5 pounds per day would supply all of the energy needs of the United States! It was the realization that quite modest amounts of matter could produce enormous amounts of energy that was one of the main motivations for developing nuclear reactors.

Conversely, if our interest is in making new kinds of particles—in studying matter that has been created artificially—then the Einstein equation tells us that it will require large amounts of energy to do so. We will see that the usual method of gaining such energy is to allow a very fast particle to collide with one at rest, and to convert the kinetic energy of the fast particle into the mass needed to make the new particles in the process. This method, of course, requires a supply of particles of very high energy—particles whose velocity is greater than 90 percent of the speed of light. In the early 1930s, when the first important results of elementary particle experiments were being gathered, the only source of particles that had energies as high as this were cosmic rays.

Cosmic Ray Experiments

DEEP inside most stars the temperatures and pressures set off nuclear reactions that, in turn, produce energy that percolates through the body of the star and prevents the star from collapsing. When this energy reaches the surface of the star, it is radiated into space. Some of this radiation is in the form of visible light, and this is what we see when we look at the star. Other portions, not visible, are in the form of radio waves, X rays, and other types of radiation that have only lately been detected and measured by astronomers. And, almost as an afterthought, the surface of the star emits streams of elementary particles, primarily protons, that move off into space. If the star happens to be our sun, these particles form what is called the solar wind—a tenuous stream of particles blowing by the earth. If the star happens to be far away, these particles may travel for millions of years without encountering any solid body. Eventually, however, some infinitesimal fraction of the particles from the sun and other stars in our galaxy will strike the upper atmosphere of the earth; when they do, we call them cosmic rays. Although the emission of cosmic rays is not a particularly important part of a star's energy budget, these particles had an enormous impact on our understanding of elementary particles.

The great majority of the cosmic rays that come to the earth are protons, although, as one might expect because of their origin as debris from star processes, there are also scatterings of all sorts of other particles. For example, entire nuclei of uranium atoms have been seen in Skylab experiments. Those cosmic rays that come from our own sun are mostly of quite modest energies; in fact, comparable to the energies of particles emitted by radioactive nuclei here on Earth. A few cosmic rays, however, have been accelerated somewhere in the Galaxy to extremely high energies by a process that we do not really understand. This small minority of high-energy cosmic rays are the ones that are of interest to us. As we shall see in later chapters, it is possible for the energies of these protons to be even higher than those that can be produced in the largest machines that have been built or are planned today.

When one of these high-energy protons enters the earth's atmosphere, it will descend until it collides with a nucleus. On the average, a cosmic ray will penetrate about a half mile into the atmosphere before such a collision occurs, although individual protons may travel shorter or longer distances than this. As a result of this collision, some of the cosmic ray's enormous store of kinetic energy will be converted into mass, and a spray of particles will be produced. Some of these secondary particles will also have a high kinetic energy, so that when they, in turn, interact with nuclei still deeper in the atmosphere, they, too, will produce sprays of secondary particles. In this way a process called a cascade develops, in which the secondary products of collisions at one level become primary projectiles and initiate collisions on the next level down. The development of a cascade appears in Illustration 17.

At each level of the cascade some particles collide with atmospheric nuclei and produce more particles, while others continue to move toward the earth's surface. When they strike the surface, they are termed a cosmic ray shower. Depending on the energy of the incoming cosmic

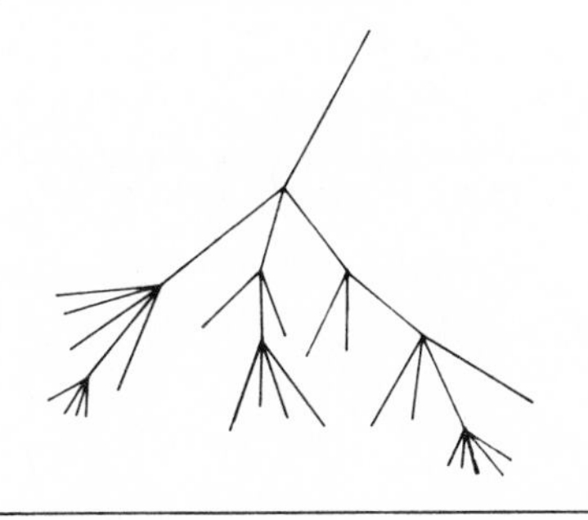

17. The development of a cascade.

ray, these showers can contain millions of particles and can extend over several square miles of the earth's surface. As you are sitting and reading this book, particles from processes like this cascade are passing through your body at the rate of several a minute.

From this discussion, it is clear that cosmic ray interactions in the atmosphere provide a sort of natural laboratory in which the interactions of elementary particles can be studied and, more particularly, in which the process of matter creation can be observed and analyzed. Before we go on to describe some of the important experiments, two of the general principles about the particle creation process should be examined.

If we took all the energy (kinetic plus mass) of all the particles created in the cascade and added to it the increased kinetic energy of molecules in the air due to the cascade, the result would be equal to the energy of the incoming cosmic ray. In other words, even though particle creation is occurring, the law of conservation of energy holds in the cascade. Furthermore, if we took any single collision in the cascade and added up the energies before and after the collision, they too would balance. In other words, the processes in the cascade may convert energy from its kinetic form to mass, but they do not violate any known law of energy conservation.

In a similar way, the electrical charge is conserved in the cascade as a whole and at each step in the cascade. For example, if the initial particle is a positively charged proton, then if there are a million negatively charged particles created during the cascade, there will have to be a million and one positively charged particles created, so that the net total charge of the cascade is always $+1$. But, with the exception of the constraints imposed by the conservation of energy and charge (and a few other conservation laws to be discussed later), anything can happen in the cascade, which means that the cosmic ray "laboratory" can provide us with a look at a wide variety of the processes that can occur when elementary particles collide with each other. This variety is one good reason why so much of our knowledge of the properties of elementary particles came originally from cosmic ray experiments.

The Discovery of the Positron: A Typical Cosmic Ray Experiment

TO illustrate the richness of the information that was gleaned by physicists in the 1930s and 1940s from cosmic ray experiments, consider one of the most important—the discovery of the first piece of antimatter by Carl D. Anderson in 1932 at the California Institute of Technology. The experiment was designed to detect and identify those portions of cosmic ray showers that reached ground level. To this end, it was necessary to do three things: (1) find out when a particle had passed through the experimental apparatus; (2) find the charge of the particle; and (3) find the mass of the particle.

"Seeing" an elementary particle is not an easy thing to do, since they are so small. If we build the right apparatus, however, we can use some natural processes to tell us what path a particle has followed. In the 1930s, the right apparatus was the Wilson cloud chamber (Illus. 18). A container with a moveable piston for one wall is filled with air that is saturated with something similar to alcohol vapor. A particle passing through this chamber will, because of its electrical charge, leave behind it a trail of atoms from which electrons have been torn. These atoms are called ions. If the piston is suddenly lowered just after the particle has left the chamber, the alcohol in the air will start to condense into drops, much as water comes out of air as dew when the temperature falls. It turns out that the ions that have been left in the wake of the particle can act as nuclei around which drops of alcohol will form, so the net result of this operation is that there will be a trail of visible drops in the otherwise hazy and diffuse cloud of condensate formed in the chamber. This trail will be located along the path that the elementary particle traversed. Although this process does not allow us to see where the

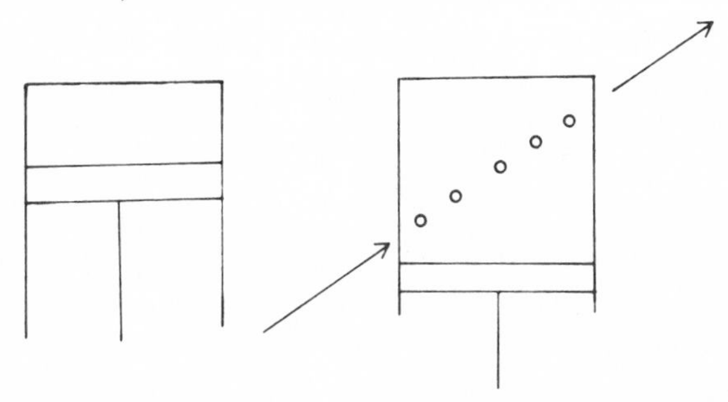

18. The Wilson cloud chamber.

particle *is,* it does allow us to see where it *was,* which for our purposes is just as good.

The presence of a track of droplets in a cloud chamber shows that a particle has been in the chamber and has been detected. In addition, it turns out that the number of drops (which is related to the number of ions created by the particle passing through the chamber) gives us some information about the velocity and mass of the particle. To determine these two variables exactly and to ascertain the electrical charge of the particle, one new feature has to be added to the experiment.

Suppose that the cloud chamber in Illustration 18 were placed between the poles of a large magnet, so that the entire chamber was inside a magnetic field. The laws of electromagnetism tell us that a particle entering the chamber from the left, as shown, will not go through the chamber in a straight line, but will follow a curved path. Hence, the droplet track will not be straight, but will be part of the arc of a circle. The amount of curvature of the track will tell us the momentum of the particle, and the direction of curvature will tell us the charge of the particle. For example, if the north pole of the magnet is above the cloud chamber and the south pole below, a negatively charged particle, such as the electron, will be bent in such a way that its path will curve upward (out of the page), while a positively charged particle's path will curve downward (into the page).

Thus, a cloud chamber located between the poles of a magnet is a good device for identifying particles. From the number of droplets and the amount of curvature of the track we can determine the mass and velocity of the particle, and from the direction of curvature we can find the electrical charge. We shall see later that the twin measurements of the ability of a particle to ionize nearby atoms and its reaction to a magnetic field are still the main criteria for the identification of particles in the laboratory.

In 1932 Carl Anderson used an apparatus that he and Robert Millikan had built at the California Institute of Technology to look at cosmic ray showers. The device consisted of a set of powerful magnets to bend the paths of particles and a cloud chamber so placed that those particles coming from a vertical or near-vertical direction would pass through the longest dimension of the chamber. In a series of papers in *Physical Review,* Anderson proved conclusively that a large number of the particles passing through his apparatus had a mass about equal to that of the electron, only they had a positive electrical charge. He called this new particle the *positron*—a contraction of positive electron—and this name has stuck. His suggestion that in view of this discovery we ought to refer to the electron as a "negatron," fortunately, was not widely adopted.

In the usual manner of experimentalists reporting a new finding, Anderson's original papers did not spend much time on speculation about the nature of the positron, but were devoted to establishing its existence beyond question. It was only later that the full import of the discovery was realized. For the first time, there was conclusive laboratory evidence that an entirely new type of matter existed—a type we now call antimatter.

Antimatter

In normal matter, electrons always have negative electrical charges and the nucleus of the atom always has a positive charge. The positron, therefore, is not a particle that "hides" in a normal atom until it is shaken loose in a cosmic ray collision. Hence, it must be one of the particles created in the cascade by the conversion of energy to matter. This process is now well understood and is seen daily in modern laboratories.

As an example, consider a high-energy photon impinging on an atom in the atmosphere (see Illus. 19). After the photon strikes the nucleus of the atom, some debris (pieces of the nucleus) will result, perhaps with some miscellaneous particles created from the energy of the photon, and, most important, an electron and a positron. We denote the former by the symbol e^- and the latter by e^+. Often, a process occurs like the

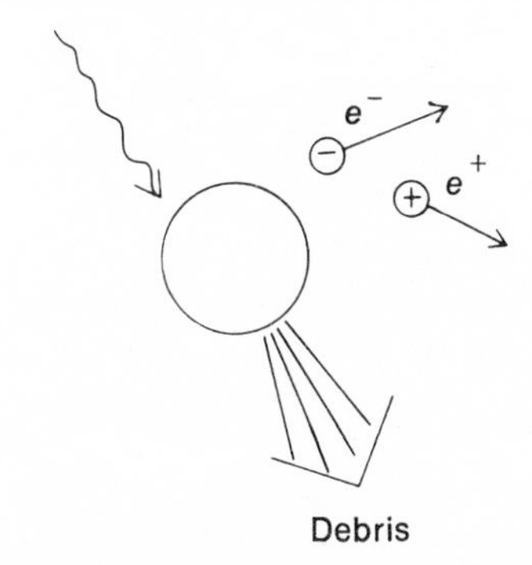

19. The production of a positron.

one sketched in Illustration 20: A photon collides with a nucleus and only the electron and positron come out. In either case, we refer to the occurrence as pair creation and speak of the electron-positron pair. It is important to realize that in this process the electron and positron are both created from energy, and that neither is part of the nuclear debris.

Energy is conserved in this process; that is, the kinetic energy of the incoming photon is converted to the mass and kinetic energies of the pair. Charge is also conserved, since every time a positively charged positron is created a negatively charged electron is created with it. This is a specific example of the general principle we discussed in the previous section—the principle that every process at the level of elementary particles is subject to the restrictions imposed by the conservation laws.

If we now allow the positron that we created to encounter an electron, a startling event takes place. The electron and positron disappear, and in their place we find some high-energy photons (see Illus. 21). This process is called *annihilation,* and it represents the inverse of creation.

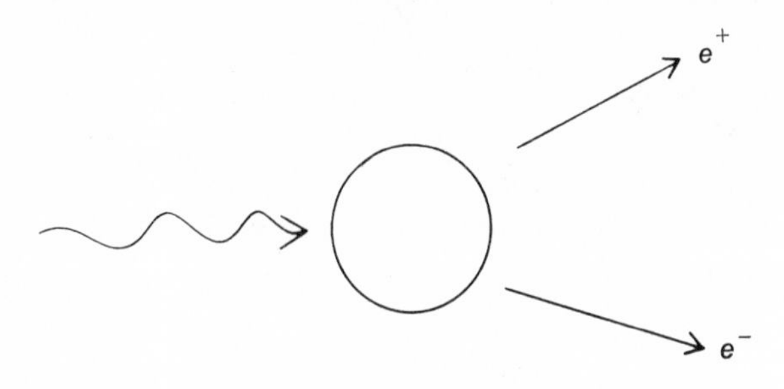

20. The process of pair creation.

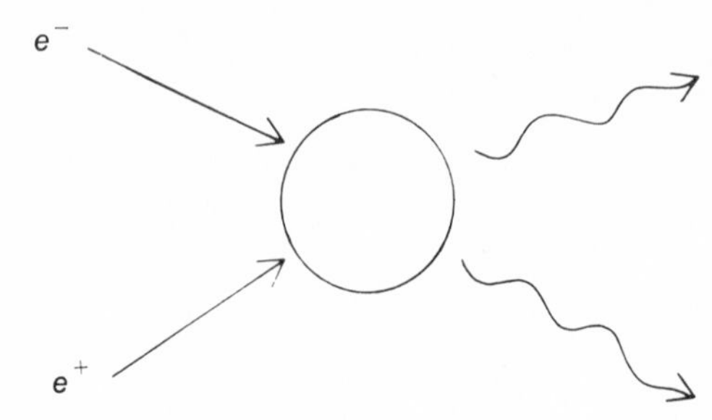

21. The annihilation of an electron-positron pair.

When a particle and its antiparticle meet, their total energy (including the energy in their masses) is converted into photons, and the original particles cease to exist.

In general, after a positron is created, it will wander until it encounters an electron with which it can annihilate. This electron could be the one with which it was created, but is much more likely to be a particle met more or less at random.

The creation and annihilation of electron-positron pairs provides a vivid and compelling verification of the principle of the equivalence of mass and energy. Since 1932, it has been found that for every particle in nature there is an antiparticle. (We will be discussing some of these discoveries later.) Actually, the existence of a particle like the positron was predicted in 1930 by the British theoretical physicist Paul A. M. Dirac. His reasoning provides an interesting way of visualizing the pair creation and annihilation processes.

Dirac did not start by thinking about antimatter, but instead was trying to resolve a rather difficult theoretical problem that had arisen when physicists had tried to combine ordinary quantum mechanics with the principles of relativity in order to get a quantum mechanical description of high-energy particles. They found that the equations predicted that particles, such as the electron, could exist in states where the energy was actually negative.

This, of course, is nonsense. If electrons had negative energy states into which they could fall, the tendency of every system in nature to reach the lowest possible level of energy would cause the electrons to start falling into these negative states. The electron's situation would be analogous to that of a stone poised on the side of a hill of infinite length. Once it started rolling, it would keep on going, falling to lower and lower energy levels as time went by. If negative energy states for electrons did exist, and if there was nothing to prevent electrons from falling into them, all of the electrons in the universe would fall into states at $-\infty$, and the radiation released when they did so would fill the universe. Consequently, physicists tended to ignore the possibility of negative energy states, or to think of them as some sort of quirk in the mathematics that would be resolved later. Dirac chose, instead, to take the possibility of negative energy states very seriously. Suppose, he said, that the states are really there, but that they are completely filled up with electrons already? In other words, if we supposed that the possible energy states of the electron are like the ones shown in Illustration 22, then the electron could have an energy mc^2 (this would correspond to an electron with no kinetic energy) or it could have some higher energy (corresponding to the electron having both mass and kinetic energy). The predicted negative states would then extend from $-mc^2$ on down to $-\infty$.

If these states are there and if they are always filled with electrons (a condition referred to as a filled negative-energy sea), two consequences follow. First, electrons cannot fall into these negative states and descend to $-\infty$ for the simple reason that there is no room for them to do so—any state they could descend into is already filled. Second, the entity we normally call the vacuum would be a filled negative-energy sea with no particles in the positive energy states.

In terms of our stone-on-the-hill analogy, Dirac's suggestion was that the stone could not start rolling down the hill because the lower slopes were already completely covered with stones.

Illustration 22 shows that pair creation works like this: A photon comes along and propels an electron from the negative-energy sea to a positive energy state. The final result will be a positive energy electron and the absence of a negative-energy negatively charged particle. If you think about the double negatives for a while, you can probably convince yourself that the absence of a negative-energy negative charge is the same as the presence of a positive-energy positive charge, and that is what we have been calling a positron. Thus, in Dirac's picture, the positron is thought of as the effect of the empty state that results when an electron is moved to a state of positive energy.

In the same way, pair annihilation occurs when a positive energy electron encounters a hole in the negative-energy sea (which, remember, is a positron) and falls into that hole. From a state of having two

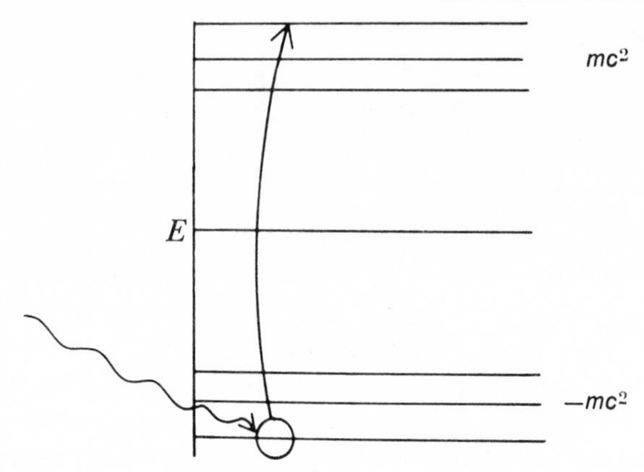

22. How pair creation works.

"particles," we wind up with a filled negative-energy sea again, with some photons moving off.

One way of visualizing the Dirac theory of the positron is to imagine a plot of level ground. If you take a shovel and dig a hole, there will be two things left—a pile of dirt and a hole (which is the absence of the pile of dirt). The former would be the electron and the latter the positron. Annihilation, in this analogy, would correspond to putting the dirt back in the hole. When this is done, you are back to level ground again.

Having understood Dirac's picture, it is important to emphasize that you do *not* see the absence of a negative energy electron in the laboratory. What Anderson saw (and what you could see in a modern laboratory) is a positively charged particle with positive energy and the same mass as the electron. The Dirac picture is just a way of interpreting this observation in terms of positive and negative energies and, at the same time, resolving a rather difficult theoretical problem.

There is another interesting thought about antimatter we should discuss. We have mentioned the fact that there is a kind of symmetry in nature between matter and antimatter in the sense that there can exist an antiparticle for every known particle. On the other hand, there is a manifest asymmetry in the world we know in that everything is made of normal matter, and antimatter is a rare and exotic species. Some physicists, most notably the Swedish theoretician Hannes Alfvén, have speculated about this apparant inconsistency in nature. If antiparticles are equivalent to particles, why are there so few of them?

If there were a region of space in which unequal numbers of particles and antiparticles were mixed, it could not remain in equilibrium for very long. The process of annihilation would go on until all of one kind of particle or the other were used up, and the result would be a region in space, similar to our own, in which one type of matter predominated. Thus, if you believe that nature really intended to have particles and antiparticles on an equal footing, you would argue that our present "normal" state is the result of the annihilation of antimatter in our region during the early history of the universe. And if we happen to live in a region that started life with an excess of particles, then there ought to be regions of space where the opposite is true; regions where, after the annihilation has occurred, only antimatter is left. In fact, there ought to be entire galaxies made of antimatter in the same way that ours is made of normal matter. In other words, it might be that some of the galaxies that we see with our telescopes are composed entirely of antimatter. There is certainly no way we can be sure just by looking at them. Light emission, as we have seen, is a process that depends on the kinds of orbits that electrons describe around a nucleus. If we replaced every electron in an atom by positrons, however, and every nucleus by an antinucleus made of antiprotons and antineutrons, then the positron orbits would be identical to the ordinary electron orbits, and the light emitted by this "antiatom" would be identical to the light emitted by normal atoms.

Of course, one way of telling if distant galaxies are antimatter would be to see what happens when material from that galaxy comes near us. It has been suggested that the Tunguska crater in Siberia is the result of an impact of an antimeteor, although there is not much scientific support for this idea. Another variation on this theme (and probably one that has more scientific merit) is to look for telltale radiation that should be produced by the annihilation process when particles and antiparticles meet at the intersections of their respective "turfs." To date, searches for this kind of radiation have not been successful, but it is safe to say that the idea that we may be looking at regions of the universe totally populated with antimatter has not been completely abandoned by modern scientists.

CHAPTER V

The Discovery of Mesons and Other Strange Things

INTERLOCUTOR: *Here's the shaggy dog you advertised as lost in the paper yesterday.*
MR. BONES: *Good Lord, it wasn't* that *shaggy.*
—"The Shaggy Dog Story,"
ANONYMOUS

The Discovery of a Meson

THE type of cloud chamber in which the positron was discovered proved to be a tremendously useful piece of equipment. As so often happens when a new detection method is developed to explore new fields, discoveries come thick and fast. The discoveries of Galileo after he built his first telescope are historical examples of this generality. Within months of the time when the telescope was first turned toward the heavens, he had seen the moons of Jupiter, mountains on the moon, and sunspots. In a similar way, when scientists first began looking at cosmic ray showers with an instrument capable of measuring both ionization and the amount of bending in a magnetic field, entirely new and unexpected results began to show up. These results were obtained during the 1930s and 1940s, and they led to the creation of an entirely new area of research—the field of elementary particle physics.

In the early 1930s, cloud chamber experiments of the type that led to the discovery of the positron were being performed in many laboratories around the world. By 1934 it had become apparent that something was wrong with the way these experiments were being interpreted. Particles were seen whose ability to create ions in the cloud chamber did not seem to correspond with the behavior of electrons, positrons, or protons. These new particles, which went by the name of "penetrating rays" (because they could penetrate through the atmosphere to the apparatus at sea level), were characterized by either positive or negative electrical charges. When the new quantum mechanics was used to calculate the progression of electrons through a cosmic ray shower, it showed that very few would be expected to reach sea level with energies like those seen. On the other hand, the energies that the particles lost through ionization in cloud chambers were too low for them to be protons. This meant either that penetrating radiation was not made up of known particles or that the quantum theory gave incorrect answers when applied to particles of very high energy. The resolution of this quandary became one of the central issues in cosmic ray physics in the late 1930s.

The fact that a problem such as this existed illustrates an important point about the way scientific research is done, a point that is often ignored when we look back at important achievements. There is a temptation to pick out the significant results and put together a chain of reasoning which, *in retrospect,* seems very precise and logical. In real life, however, the scientist is confronted with evidence that is ambiguous. In this case, for example, the discrepancy could arise because the theory was wrong (after all, it was only in 1936 that the quantum mechanical theory for electron showers was worked out). On the other hand, the technical details of the analysis of cloud chamber pictures made it very difficult to determine the mass of particles going through at high speed. Such particles would be only slightly bent by the magnets, and the normal experiemental errors could easily mask significant results. Only when all these ambiguities had been eliminated could the conclusion be drawn that particles were being seen that had a mass different from both the proton and the electron.

By 1938 enough experimental evidence had accumulated to convince physicists that the explanation of penetrating radiation did not lie in an error in quantum mechanics. In an effort to give a final resolution to the problem, Seth Neddermeyer and Carl Anderson, working with the California Institute of Technology cloud chamber, tried a new mode of operation. They inserted a Geiger counter into the chamber, and then arranged to expand it and photograph the drops only when a particular reading was seen in the counter, a reading that indicated that a very slowly moving particle was in the chamber. In this way, they hoped to obtain a series of photographs of the tracks of penetrating rays, which would be easy to analyze. Luck was with them and their program ultimately proved successful. In a letter to *Physical Review* on June 6, 1938, they reported a single event in which a penetrating ray of positive charge was slowed down enough by passing through the Geiger counter so that it came to a complete stop in the cloud chamber. It turns out that this was the best possible experimental situation for determining the mass of a particle, and when Neddermeyer and Anderson analyzed the photograph, they reported that a new particle had been discovered, with a mass about 240 times that of the electron. (The modern value for this number is 210.) They named their new discovery the "mesotron," from the Greek root "meso" (which means middle). This was later shortened to *meson,* a term that is both more convenient and more correct linguistically. This particle was customarily denoted by the Greek letter μ (mu); it is therefore now called the mu-meson, or muon.

There were actually two mu-mesons, one with a positive charge and one with a negative, and the masses of the two were identical. Were these, then, the particles that Yukawa had predicted would be responsible for the strong force? Again, while it is tempting to think that science proceeds in this straightforward way, in point of fact there had been many predictions of intermediate mass particles. By 1938, Yukawa's was only one, and the suggestion, coming as it did from a research center far away, was not even the most prominent one in the minds of American scientists. Thus, the question that had to be answered was whether this meson behaved as one would expect if it were really the "nuclear glue" we previously mentioned.

The first thing that became obvious was that the muon was not stable in the sense that the electron and proton are. Like the neutron, it decays in a certain amount of time into other particles. The lifetime of the muon is measured to be about 10^{-6} second, and it decays by the process $\mu \rightarrow e +$ two neutrinos. Of course, the two neutrinos cannot be seen in the cloud chamber, since they are uncharged and therefore do not create ions.

When a mu-meson enters a block of material, it will encounter the atoms that make up that material. If the mu-meson is indeed the particle responsible for holding the nucleus together, we would expect that when it came near a nucleus it would interact strongly with it. We know that a typical time scale for strong interactions is approximately 10^{-24} second, and since this is much shorter than the lifetime of the muon, we would expect that most of the muons that are slowed down and stopped in a block of material would wind up interacting with a nucleus long before they had a chance to decay. This, in turn, means that if we look at what comes out of the other side of such a block of material, we should not see the characteristic electrons that result from muon decay, but rather the kind of nuclear debris that is associated with nuclear reactions themselves.

Actually, there is a slight quibble to be made about this argument. When the theory of the process by which mu-mesons are captured in atoms was worked out, it appeared that the positively charged muon would have to be repelled from the nucleus by the ordinary electrical force, so that it would decay normally without entering a nucleus. The argument given above, however, held for the muon with negative charge, since it would be attracted toward the nucleus by the electrical force. This quibble lost its importance, however, when experiments showed that decay electrons were seen for *both* kinds of muons when they entered a block of material. Both kinds were somehow staying alive for the full 10^{-6} second required for them to decay.

Thus, by 1947 physicists were faced with a real dilemma. The mesons were supposed to be the particles that bound the nucleus together, yet when a meson came near a nucleus it showed no inclination to interact strongly with it. If the meson *inside* the nucleus interacted strongly enough with protons and neutrons to overcome the electrostatic repulsion, how could it interact so weakly when it was *outside* the nucleus? Physicist I. I. Rabi of Columbia expressed the sentiments of the physics community about the mu-meson very well with the query, "Who ordered this?" The predicted meson had been found all right, but it turned out to be the wrong one!

The Pi-Meson: One Dilemma Resolved, Another Created

WE have seen that one of the major experimental problems that arises in the study of elementary particles is finding ways of detecting the presence of the particles. The cloud chamber solved this problem by using ions created by the particle as condensation centers for alcohol drops. During the period immediately after World War II, a similar technique using photographic emulsions came into widespread use in cosmic ray studies.

In ordinary photographic film, light striking the emulsion causes a chemical reaction which, when the film is developed, leads to grains of silver being deposited on the negative. These grains are opaque, so that after development the film will be dark where light was present and lighter where it was not. In this way, the characteristic reversed negative of a photograph is formed.

A very similar process can be used to detect the presence of a charged particle. When such a particle passes through a specially prepared photographic emulsion, it causes reactions which, upon development, lead to silver grains being deposited along the path that the particle followed. Someone examining the emulsion with a microscope will then be able to see where the particle had been by following the trail of these grains, which show up as dark spots on the lighter background.

This new technique had several advantages over the cloud chamber. A stack of emulsion plates could be left on a mountaintop for months at a time, and every particle that passed through them would be recorded. In this way, the emulsion could "see" many more particles than could a cloud chamber that was operated sporadically. In addition, the small size of the silver grains made it possible to see particles in the emulsion even if they only traveled 10^{-6} centimeter, something that could not be done when the detection depended on the formation of droplets. Finally, and most important, the photographic emulsion was much denser than the air-alcohol mixture used in a cloud chamber, so that particles produced by collisions in the emulsion were likely to encounter another nucleus and interact with it before they could leave. In this way, the emulsion could be used as a target in which new particles were created, and, at the same time, as a detector that measured the way the new particles interacted with nuclei.

In 1948 a group of physicists headed by Cecil F. Powell at the University of Bristol in England began publishing the results of their examination of emulsions that had been exposed at the height of 10,000 feet on the Pic du Midi in the French Alps and on other high mountains around the world. They saw tracks of the mu-mesons, of course, but, in addition, they saw collisions of energetic particles with nuclei that produced another kind of meson that was heavier than the muon. Once this heavier meson was produced, one of two things would happen. The new meson could decay in about 10^{-8} second into a muon and an uncharged particle (which was assumed to be a neutrino), or it would interact with another nucleus in the emulsion. When it did the latter, the evidence of the tracks in the emulsion showed that the new meson interacted strongly, breaking up the nucleus and creating a spray of debris. In other words, this particular meson seemed to be the one that Yukawa had been talking about. It not only had a mass intermediate between that of the proton and the electron, but it interacted strongly when it came near a nucleus, something that the mu-meson did not seem to do.

The new particle was christened with the Greek letter π (pi) and called the pi-meson, or pion. Because of the possibility that the new pi-meson played a fundamental role in binding the nucleus together, it quickly became an object of intense study throughout the physics community. In Chapter VI we will talk about the development of particle accelerators—machines that can take protons (or electrons) and accelerate them to energies comparable to those of cosmic rays. In the early 1950s machines were becoming available that were capable of creating pions for use in physics research, so that detailed studies of their properties could be done. It turned out that the meson come in three varieties —there are pions with a positive electrical charge, pions with a negative electrical charge, and pions that are electrically neutral. These are denoted, respectively, by the symbols π^+, π^-, and π°. The mass of the charged meson is 273 times that of the electrons, and each of the charged mesons decays via the reaction $\pi \rightarrow \mu + \nu$, with a lifetime of about 10^{-8} second. The π° has a mass 265 times that of the electron and decays via the reaction $\pi^\circ \rightarrow$ 2 photons in about 10^{-16} second. It is now believed that most of the strong force is generated by the exchange of mesons in the nucleus, as we discussed in Chapter IV. The pi-meson is, therefore, an extremely important addition to the ranks of known particles.

Why was it not discovered sooner? Given the importance of the meson in the theory of the nuclear force, why did it take a decade to unravel the π-μ problem?

Part of the answer to this question is historical—the decade between the discovery of the muon and the identification of the pion with the Yukawa meson spans World War II, a period when most physicists were focusing their attention on more pressing questions. But perhaps more important are the properties of the pi-meson itself. When such a meson is created by a cosmic ray collision high in the atmosphere, one of two things can happen: It can decay before it hits the ground or it can interact with a nucleus in the atmosphere. Typically, a pi-meson of moderate energy will travel only a few meters or tens of meters before it decays, and even if the energy were high enough to bring the meson to sea level before decay, it would be able to travel only a few hundred meters through the atmosphere before interacting with a nucleus. In either case, the original meson will not reach ground level and therefore would not be seen in a ground-based cloud chamber experiment. And since most pi-mesons in a cosmic ray shower are created at high altitudes, before the energy of the particles is degraded below the level needed for particle production, pi-mesons are never seen at sea level. This fact explains, incidentally, why the Bristol group found evidence for the pion when they exposed emulsions on a mountaintop rather than at sea level. It also explains why, in the acknowledgments of one of the original pi-meson papers, the authors thank the leader of a mountaineering expedition for carrying some plates to the top of Mount Kilimanjaro in Tanzania—a height of 19,000 feet.

But even a cloud chamber on a mountaintop would not stand a very good chance of actually detecting and identifying a pi-meson. As we have discussed, the material in a cloud chamber is not very dense, so that a meson passing through the chamber has a relatively low probability of encountering a nucleus and interacting. It would, therefore, be very difficult to tell the difference between a fast muon and a fast pion by looking at droplets. We have already seen how critical it is for a particle to be brought to rest in a cloud chamber in order for it to be identified, and how the first identification of the mu-meson depended on a lucky event in a ground-level experiment. Given the difficulties attendant on operating cloud chambers on top of a mountain peak and the resulting paucity of high-altitude cloud chamber data, it is not too surprising that there was no equivalent lucky event for the pi-meson.

With emulsions, however, the situation was different. They are quite dense, so that pions entering the emulsion are likely to be stopped. In addition, it turns out that it is possible to make much more accurate determinations of particle mass by counting silver grains through a microscope than by analyzing droplets.

By 1948, therefore, the riddle of the mesons had been solved. Not one, but two groups of particles with a mass between that of the electron and the proton had been found. The pions are the particles predicted by Yukawa, and all three of them are routinely exchanged within the nucleus to generate the strong interaction. In a sense, once Rutherford had discovered the nucleus, the existence of such a particle was inevitable. Thus, the number of elementary particles is increased by one (it is customary to refer to all of the pion family as a single particle). In return for this complication in our picture of the universe, we gain an understanding of the strong interaction.

The case of the mu-meson is not so clear. It adds another elementary particle to our fast-growing collection, but it is not obvious just what role it plays. It is not essential to our understanding of the nucleus, and in many ways it seems as if nature, having created the electron, went ahead and repeated the process for a particle 200 times heavier. One of the major unsolved mysteries of particle physics remains the question of why the muon should exist at all or, in the words of Nobel Laureate Richard Feynman, "Why does the muon weigh?"

Stranger Still

AT the same time that the evidence for the pi-meson was accumulating on photographic plates on mountaintops, two researchers at Manchester University in England began reporting some very unusual events in their cloud chamber photographs. We have seen that one of the drawbacks of the cloud chamber is the fact that the air-alcohol mixture in the chamber has such a low density. To get around this difficulty, experimenters began inserting plates of heavy materials, such as lead, into the chamber to slow the particles. Pictures similar to Illustration 23, below, would then be seen. A particle would enter the chamber from the top and go into the lead plate. It would strike a lead nucleus, and the debris of this collision would then be seen emerging from the other side of the plate.

In December 1947, however, a rather unusual event was reported. It is shown in the right-hand portion of the illustration. The usual particle-above–debris-below pattern was seen, but in addition a set of V-

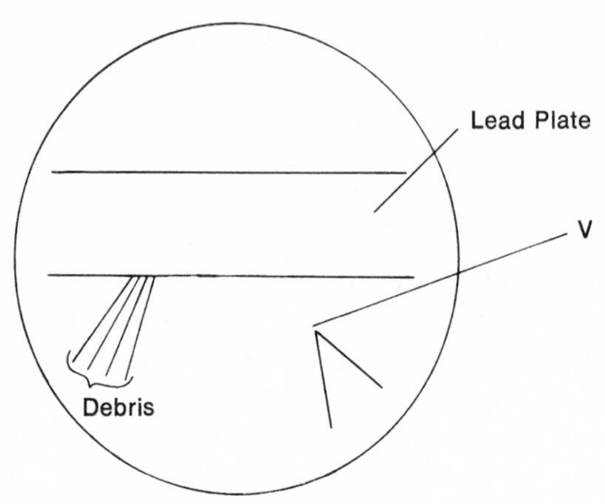

23. Discovery of the lambda particle.

shaped tracks seemed to appear from nowhere on the far side of the plate. The only possible interpretation of this event was that an uncharged particle was created in the lead plate. Such a particle would not create ions and would not be revealed by the droplets in the chamber. At the point labeled **V** this uncharged particle then decayed into two charged particles, which were visible in the chamber. Later work showed that these two charged particles were, in fact, a proton and a negative pi-meson. In modern terminology, the uncharged particle is called the Λ (lambda) particle, and is customarily written Λ° to emphasize its lack of electrical charge. Thus, the reaction sketched in Illustration 23 would be $\Lambda^\circ \rightarrow p + \pi^-$.

There are a number of extraordinary things about this event. In the first place, the fact that the lambda can decay into a proton plus something else clearly implies that it must have a mass greater than that of the proton. No one had expected or anticipated that such a particle could exist. Even more surprising, the fact that the lambda seemed to travel several centimeters in the chamber before it decayed indicated that it had a very long lifetime. We can get a rough estimate by noting that the time it takes light to travel 3 centimeters is

$$t \approx \frac{3}{3 \times 10^{10}} = 10^{-10} \text{ sec}$$

so that the lifetime of the particle would have to be of this order of magnitude in order for its tracks to be seen in the cloud chamber. In fact, the lifetime of the lambda is now known to be 2.6×10^{-10} second. While this may seem like a very short time on the human scale of things, it is very long when compared to the 10^{-24} second that we saw was "natural" for particles participating in the strong interaction.

This fact poses a serious problem for our understanding of the nature of the lambda. Physicists had been able to accept the relatively long lifetimes of the pi- and mu-mesons because they decayed by a process something like the weak decay of the neutron. The fact that the end products of pion and muon decay are particles that do not participate in the strong interaction made it easy to accept the fact that, on the nuclear scale, they lived a long time. With the lambda, however, this argument cannot be made. The lambda is created in a strong interaction (the disintegration of a nucleus) and it decays into particles that manifestly participate in the strong interaction. By reasonable analogy, the lambda *ought* to undergo decay in something like 10^{-24} second rather than having a decay time that is more characteristic of weak interactions. But it does not.

This property of the lambda earned for it an unusual name. Together with a few other particles that have similar properties, it was dubbed a "strange" particle. As we shall see later, the quantity that physicists call "strangeness" can be given a quantitative meaning (contrary to what you might suppose), but for the moment let us use it merely as a label for those particles that seem to decay much more slowly than expected.

At about the same time as the discovery of the lambda, a group of strange mesons was detected in cosmic ray experiments. They are now called the K-mesons, and they come in two pairs. The first pair contains a positively charged meson called the K^+ and a neutral member called the K°. The second pair contains a negatively charged member called the K^- and a neutral member that is the antiparticle of the K° and is called the $\overline{K^\circ}$. All of these particles have a mass about 1,000 times that of the electron.

With the discovery of strange particles in cosmic ray experiments, the whole nature of elementary particle physics changed. From the study of the few known particles and a search for a few predicted ones, scientists turned to the much more general question of how many particles could actually be produced and what their properties were. To carry out this sort of study, the kind of cosmic ray experiment we have been discussing is really not suitable. Having to wait and hope to see a favorable event may be a good way to map the general features of a new field, but when exhaustive studies need to be made, it becomes necessary to have a source of energetic particles that can be controlled. With such a source, it should be possible to create the new particles at will so they can be studied in much more detail than is possible with cosmic rays.

Fortunately, during the 1930s machines capable of producing such energetic beams were being developed. They are called accelerators and bear such exotic sounding names as cyclotron, synchrotron, and linac. We will talk about the development of these machines and the discoveries made with them in Chapter VI.

A Question of Units

UP to this point, we have talked about the masses of the elementary particles either in terms of grams or in relation to the mass of the electron. Once a serious study of these particles starts, however, it is pretty clear that neither of these two sets of units is particularly useful. For scientists who used them all the time, writing all of those 10^{-34}'s soon became tiresome, and there seemed to be little point in referring everything to the mass of the electron in processes in which that particle was not involved. The system of units now used in discussing particles is based on the equivalence of mass and energy, and has as its basic unit the electron volt, written eV.

The electron volt is defined as the amount of energy gained by a particle whose charge is equal to that of the electron when it moves through a 1 volt potential difference. For example, a single electron that moves from one pole of an ordinary car battery to another would acquire 12 electron volts of energy. Ten electrons following the same route would acquire a total of 120 electron volts of energy, as would a single electron moving across a larger battery rated at 120 volts. The electron gains energy in such a process because it takes work to move it from one side of the battery to the other against the electrical force. In the system where mass is measured in grams and length in centimeters, 1 electron volt $= 1.6 \times 10^{-12}$ erg.

Since energy and mass are related by the Einstein equation, we can talk about the energy equivalent of the mass of the electron. In terms of electron volts, we have for the electron

$$m_e c^2 = 0.51 \times 10^6 \text{ eV}$$

while for the proton we have

$$m_p c^2 = 939 \times 10^6 \text{ eV}$$

From these two numbers, we see that the masses of elementary particles turn out to be relatively large numbers when expressed in terms of electron volts. Consequently, an abbreviation scheme is normally used, as indicated in the following table:

ABBREVIATION	FULL NAME	VALUE
eV	electron volt	1
keV	kiloelectron volt	10^3 eV
MeV	million electron volt	10^6 eV
GeV	gigaelectron volt	10^9 eV
TeV	teraelectron volt	10^{12} eV

The prefix *giga,* to designate 10^9, arises because the term "billions" is given a different meaning on the two sides of the Atlantic. To an American, a billion is a thousand million (10^9). To a European, however, it is a million million (10^{12}). For a while, the unit BeV (billion electron volts) was in wide usage, but it caused enough confusion to generate the set of prefixes in the table above. These are now used universally by international agreement.

In terms of energy units, the masses of all of the particles we have discussed so far are given in the following table:

PARTICLE	MASS (for particles) OR ENERGY (for photons)
Electron	0.511 MeV
Muon	105.7 MeV
$\pi^{\pm}$	139.6 MeV
π°	135.0 MeV
K	493.7 MeV
Proton	938.3 MeV
Neutron	939.6 MeV
Λ°	1,115.6 MeV
Photons (visible light)	~ 10 eV
X rays	~ 100 keV–10 MeV

CHAPTER VI

The Coming of the Accelerators

Silently we went round and round . . .

—OSCAR WILDE,
"The Ballad of Reading Gaol"

From Natural to Man-Made

ALL the experiments we have described up to now have involved the use of natural sources of energetic particles to trigger the events being studied. Rutherford, for example, got his alpha particle projectiles from pieces of naturally radioactive material, and Anderson discovered the positron as an end product of the collisions of cosmic rays. There are obvious advantages to this kind of experiment. They are relatively cheap, since you only have to build equipment to detect interactions: the projectiles are free. On the other hand, as we pointed out in Chapter V, there are serious limitations as well. If you depend on the natural supply of particles for an experiment, then you have no choice but to wait until the particles you want happen along.

In addition, the cosmic ray experiments revealed that there were many previously unsuspected particles in nature, and that all these particles are unstable. If we want to see what these particles are like, we must have some way of producing them in large enough quantities to study. From the fact that many of the cosmic ray discoveries involved an element of luck it was clear that any systematic work on the elementary particles would also have to wait until "lucky" events could be produced routinely in the laboratory. This, in turn, would depend on the ability of physicists to produce large quantities of energetic particles. With these "artificial cosmic rays," experiments such as the ones we have been describing could be carried out under controlled conditions.

To take a normal particle and give it high energy requires that the particle be accelerated. Machines that do this job are called accelerators (terms such as *atom smasher* never really appealed to physicists). While cosmic rays can supply us with high-energy protons, accelerators can be (and are) used to accelerate any charged particle, from protons and electrons to the nuclei of heavy atoms. By and large, however, it has been machines that accelerate protons and electrons—the most abundant and stable of the elementary particles—that have occupied the forefront of modern research.

There are two general classes of accelerators. In one kind, particles are accelerated while they travel down a long straight tube. This is a linear accelerator. In the other type, the particle is made to move in a circular path by applying a magnetic field and then boosting its energy each time it comes past a given spot on the circle. This sort of machine is called a cyclotron or a synchrotron, depending on how the magnetic field is applied. Within each of these general categories are variations and adaptations which, in the final analysis, are limited only by the ingenuity of the designers. Both protons and electrons can be (and are) accelerated in linear and circular machines.

Historically, a conference held in Bagnères, France, in July 1953 is thought of as the point at which the main research work on elementary particles shifted from cosmic rays to accelerators. Before such an event could occur, though, there had to be a long period of development during which these machines were transformed from experimental ventures into reliable tools that could be used in daily research.

E. O. Lawrence and the Cyclotron

ON a California evening in 1929, Ernest O. Lawrence, at the time an assistant professor of physics at Berkeley, sat in the university library catching up on the technical journals. In the German journal *Arkiv für*

Electrotechnik, he came across an article devoted to a scheme for producing accelerated particles. This started him thinking of another way of accomplishing the same task, and he jotted down a few notes.

Basically, his idea went like this. It had already been seen that a charged particle will be deflected by a magnetic field; in fact, if the field covers a large enough region of space, the particle will move in a circle. If the magnetic field has a strength denoted by B, then it will exert a force equal to Bqv on a particle of charge q moving with velocity v. In order to keep the particle moving in a circle of radius R, the magnetic force has to balance the centrifugal force. The formula for the latter is mv^2/R, where m is the mass of the particle. This means that for a particle in a magnetic field we must have

$$\frac{mv^2}{R} = Bqv$$

so that the radius of the circle in which the particle moves is just

$$R = \frac{mv}{Bq}$$

We can visualize the meaning of this equation from Illustration 24. If we imagine a large magnet with one pole above the page and one below it, then to each velocity that a given type of particle can have, there is

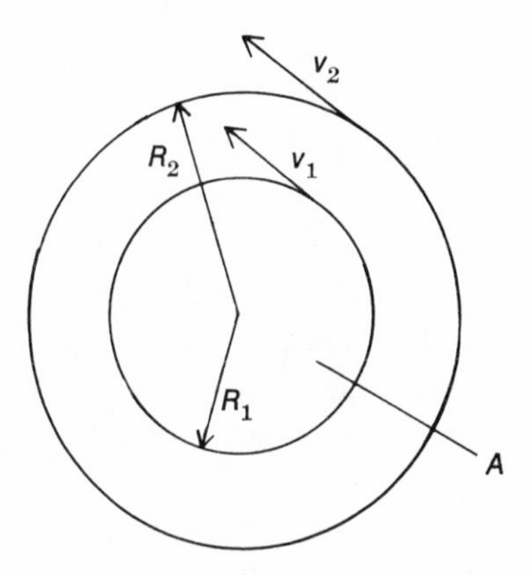

24. Cyclotron orbits. The inner orbit is what we have called orbit 1, while the outer is orbit 2.

one radius for its motion. For example, a particle moving with velocity v_1 would move in a circle of radius R_1, a particle of velocity v_2 would move in a circle of radius R_2, and so on. The faster the particle moves, the larger its radius.

These circular paths have come to be called *cyclotron orbits.* If the orbits in Illustration 24 are for protons, then electrons in the same magnetic field will move in the opposite (clockwise) direction through circles of much smaller radius. The fact that the electron radius is smaller is simply a consequence of the fact that the orbit depends on the mass of the particle, and the electron has a smaller mass than the proton.

The next question is how long it takes for a particle to get around a cyclotron orbit. In the orbit labeled 1 in the illustration, the particle is moving with a speed v_1 and has to travel a distance $2\pi R_1$, which is the circumference of the circle. This means that

$$2\pi R_1 = v_1 t$$

If we substitute for v_1 from equation $mv^2/R = Bqv$, this equation becomes

$$2\pi R_1 = \frac{Bq}{m} R_1 \cdot t$$

This equation contains the basic insight that allowed Lawrence to build the first cyclotron. Notice that the radius of the orbit, R_1, cancels out on both sides of this equation—

$$2\pi \cancel{R}_1 = \frac{Bq}{m} \cancel{R}_1 \cdot t$$

—so that the time it takes to complete an orbit depends only on the magnetic field and the charge and mass of the particle. It does *not* depend on how fast the particle is moving. The reason for this lies in the fact that faster moving particles go in larger orbits, so the increased speed of the particle is exactly canceled by the larger distance it has to travel. The time remains exactly the same.

The significance of the cancellation is this: If we installed an accelerating device in the magnetic field in the figure at the line A and waited until the particle in orbit 1 came around to that line, we could accelerate the particle. In fact, if we timed the pushes just right, we could arrange to give the particle a boost each time it came around to line *A*. In this way, we could imagine giving the particle a great deal of energy in small doses, just as one can get a child's swing going to great heights by a series of small, properly timed pushes. The only problem would be that as the particle went faster, it would move to an orbit of greater radius. And that is where the cancellation becomes important, because the pushes that are timed to a particle in orbit 1 will also be properly timed for orbit 2, or any other orbit. This means that if we start with a particle in orbit 1 and time our accelerations to give it energy, the accelerations will continue to be properly timed as the particle gains energy and moves to higher orbits. In this way, it should be possible to accelerate the particle to a very high energy by supplying a series of small appropriately timed voltages, rather than a single large one.

According to people at Berkeley at the time, when Lawrence realized the implications of this fact, he raced around the laboratory like a modern-day Archimedes, except that instead of shouting "Eureka," he kept stopping people and telling them that "*R* cancels *R*! *R* cancels *R*!"

The machine that Lawrence and his co-workers eventually developed is known as the cyclotron. The name itself started as something of a laboratory joke at Berkeley. The machine consisted of two semicircular, D-shaped magnets (called dees) arranged as shown in Illustration 25. There was a gap between these magnets across which a voltage could be applied. This voltage changes sign periodically, much as ordinary household voltage changes sign 60 times each second.

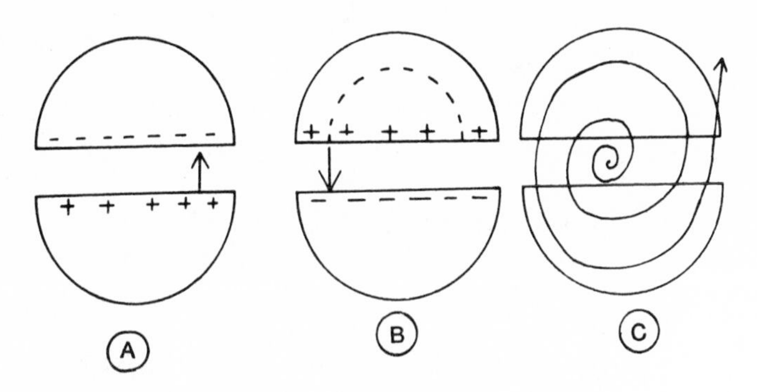

25. The principle of the cyclotron.

A proton that arrives at the right-hand gap when the far side of the gap is at a negative voltage with respect to the near side will be attracted toward the far side and pulled across the gap, acquiring energy in the process (part A). While the proton is moving through the magnetic field of the upper dee, the voltage across the gap is changing so that when the proton arrives at the left-hand gap it is again attracted by the negative voltage of the far side (part B). Once more it is accelerated, and the process repeats.

A particle injected near the center of the magnets will move in a spiral path like the one shown in part C. Each time the particle crosses a gap, it is accelerated and moved to a higher orbit until it is extracted at the outer edge of the machine. The end result of this process is a beam of energetic particles that can be used in place of cosmic rays in experiments.

The development of the cyclotron in the 1930s makes a fascinating story. Lawrence commandeered an old shack on the Berkeley campus for his laboratory and went around knocking on the doors of private foundations to raise money for the project. Meanwhile, starting with a pillbox-shaped machine 4 inches in diameter, he and a former student, Stanley Livingston, worked day and night to overcome the technical difficulties involved (some of which were truly prodigious). An improved version of the 4-inch machine yielded protons with energies of 80 keV in 1931, and in February 1932 an 11-inch version of the machine achieved Lawrence's original goal of 1 million volts (1 MeV). Livingston's recollection of the event is vivid: "I wrote the figure on the board. Lawrence came in late. . . . He saw the board, looked at the microammeter to check the resonance current, and literally danced around the room."*

The figure of 1 million volts was chosen as a goal for a number of reasons. It is a nice round number, and as such it undoubtedly made the fund raising easier. It was also generally accepted at the time that energies of this magnitude were needed to penetrate the nucleus, so Lawrence felt that an energy of at least an MeV would have to be achieved before any significant experiments could be done. People in the laboratory were so confident that this was the case that they had

*As quoted in Nuel Pharr Davis, *Lawrence and Oppenheimer* (New York: Simon and Schuster, 1968).

wired the Geiger counter, which monitored the radiation from the beam target, into the same circuit as the machine, so that it operated only when the beam was on.

Imagine their surprise, therefore, when they learned that a group of English scientists, using an accelerator of the old-fashioned design, had produced a collision in which the target nucleus had been broken apart, producing new chemical elements in the process. Realizing that if a beam of a few hundred keV could produce an artificial transmutation of elements, then their cyclotron would have been doing the same thing, Lawrence's group went into the laboratory and rewired the circuits so that the Geiger counter would stay on after the machine was turned off. Sure enough, the counter kept clicking, indicating that the cyclotron beam had produced new chemical elements that were now decaying like any other radioactive material. There must have been some pretty sick faces in the laboratory on that day. It is bad enough to be beaten to an important discovery, but to be beaten because you did not leave a machine on . . . In the words of one of the participants at that scene, "We felt like kicking each other's butts."*

But if this achievement eluded Lawrence's new radiation laboratory, other honors came thick and fast. In 1933, at the age of thirty-two, he was elected to the National Academy of Sciences as its youngest member (and first South Dakotan). In 1939 he received the Nobel Prize. The cyclotron quickly developed into the principal instrument for the study of nuclear physics, and it also became one of the prime sources for exotic radioactive materials, which were used in medicine as diagnostic tracers and for the treatment of cancer. In 1940 Edwin M. McMillan identified the elements neptunium and plutonium in targets that had been irradiated by the cyclotron beam, an achievement for which he later received the Nobel Prize. These elements do not occur in nature, but are the first such artificial chemical elements ever produced—the first of a dozen or so that are now known, and most of which were created at Berkeley.

With his brother John (an M.D.), Lawrence was quick to exploit the medical uses to which his machine could be put. In addition to the production of radium and other radioactive elements for cancer therapy, experiments were undertaken to assess the usefulness of the cyclo-

*Nuel Pharr Davis, ibid.

tron beam in the direct treatment of tumors. This is still a field in which beams of accelerated particles are used, and there are treatment facilities in the United States where beams of protons, neutrons, and even pi-mesons are used in cancer treatment.

This aspect of the cyclotron must have acquired a very special significance for the Lawrence brothers when their mother was diagnosed as having terminal cancer in 1938. In what must be one of the most dramatic and little-known episodes in the history of physics, the brothers brought her to the experimental treatment facilities in Berkeley and treated her with large amounts of radium. She may even have been the first person to have been treated with the neutron beam from the cyclotron, although there is no official evidence of it. Whatever the treatment was, it was successful, and she lived a vigorous life until the age of 83.

In many ways, Lawrence was the prototype of a figure that has become rather common in modern physics—the big-time operator. Most of the experiments that we have discussed previously have been rather modest in scope. They required relatively little in the way of financial support and could be run by a research team consisting of a scientist and a few students. Developing the cyclotron could not be done in this way. It required a major laboratory with engineers, technicians, and scientists from many different areas of specialization. It is probably the first example of what we now call team science. The man who can run such an operation needs to have abilities beyond those we normally associate with the scientist. He has to be able to "shake the money tree," coordinate the work of many different individuals, and, in his spare time, think up good experiments for his team to perform. Today, when even quite ordinary experiments at modern accelerators require dozens of workers and millions of dollars, these skills are even more in demand.

The Synchrotron

As Lawrence saw things, there was no limit to the amount of energy that a particle could be given in a cyclotron. Just make the machine bigger, he said, and we can go on accelerating forever. Just before World War II, he even managed to obtain funds for a machine that would achieve 100 MeV.

There are, however, some fundamental limits on cyclotron operation discovered through the study of the theory of relativity, one of which is that as particles start to move at speeds approximating that of light, they become more massive. Since the velocity of the particle in a cyclotron depends on the mass, this places limits on the energy of the particles in the beam. The theoretical limit is about 25 MeV, well below Lawrence's goal of 100 MeV. In fact, the highest energy cyclotron in use today produces a beam of about 22 MeV protons.

How important is this limit? Twenty MeV is an energy that is large enough to do almost any experiment involving the nucleus of the atom. But if we want to study elementary particles, it is a rather small energy. For example, to produce a pi-meson in a collision between the beam and the target, we need an absolute minimum of 140 MeV, the energy equivalent of the pion mass. Clearly, this sort of experiment cannot be done with a 20 MeV machine.

In fact, if we want to talk about accelerators that could be used to produce and study elementary particles, it is probably a good guess that we would need energies in the range of a few GeV, rather than in the range of hundreds of MeV. It follows that we would need a machine that works on an entirely new principle. This new principle was realized in the synchrotron, the basic prototype of the present generation of accelerators.

The basic limitation of the cyclotron is that it was designed to contain ever faster moving particles with a single, constant magnetic field. This difficulty was overcome in the synchrotron by increasing the magnetic field as the particle becomes more energetic. A typical synchrotron apparatus, shown in Illustration 26, below, consists of a series of magnets in the shape of a hollow ring and one (or more) places where forces can be applied to accelerate the particle. Suppose that the field in the magnets is adjusted so that a given particle is moving in a circle whose radius is exactly that of the ring itself. If this particle were not accelerated, it would simply continue to move around the ring. If it is accelerated, then the same law that applied in our discussion of the cyclotron says that the particle will move to an orbit of larger radius. If nothing were done to counteract this tendency, the particle would soon strike the wall of the machine and be lost.

Suppose, however, that we arrange things so that as soon as the particle passes the acceleration point, the field in the magnets is increased. There will then be two competing effects: The acceleration

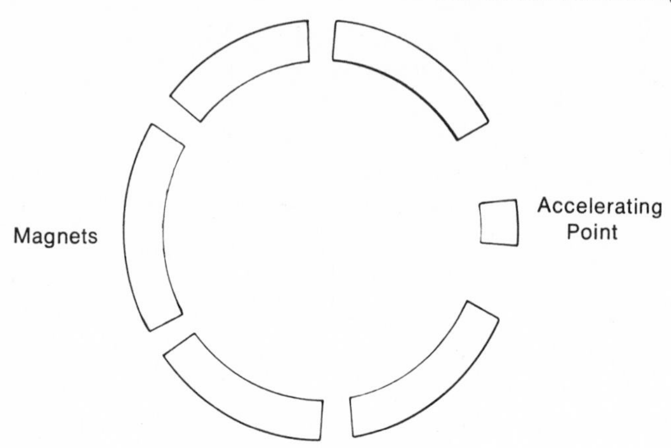

26. A typical synchrotron apparatus.

will tend to make the particle move to a larger radius, but the increased magnetic field will tend to move it to a smaller one. If we adjust the acceleration and the magnets just right, we can make these two effects cancel, so that the particle will keep moving around the ring, even though it is more energetic. Thus, by constantly stepping up the magnetic field as the particle's speed increases (a process that accelerator people call *ramping*), we can keep increasing the energy of the particle in small increments until it reaches the energy we need for a particular experiment. The only limit on the synchrotron is the size of the ring and the expense of building large machines.

Unlike the cyclotron, which can deliver a continuous beam of accelerated particles, the synchrotron runs through the ramping cycle for a bunch of particles and then starts the process again for the next bunch. It consequently delivers its accelerated particles in short bursts, rather than continuously, but this is a small price to pay for the higher energies it can attain.

The first synchrotron to break into the GeV range was built at Brookhaven National Laboratory on Long Island, New York. In 1953 it delivered a beam of 3 GeV protons from a ring 60 feet in diameter. The largest proton synchrotron is located at the Fermi National Accelerator Laboratory (Fermilab, FNAL, or, to old-timers, NAL) near Chicago (we will discuss it in more detail later). It delivers 500 GeV protons from a ring over a mile in diameter, and there are plans to upgrade it to 1 TeV.

Thus, we have come quite a way from the first 4-inch cyclotron that Lawrence put together in his laboratory.

Linear Accelerators

THE synchrotron is the machine of choice for delivering protons at very high energies. For electrons, though, it has a fundamental limitation. In Chapter I we saw that any accelerated electrical charge will emit photons. Electrons moving around in a ring are being accelerated and will therefore produce radiation. This radiation results from the forces associated with the magnetic field (i.e., with the forces that keep the electron moving in a circle), and not primarily from the modest boost the electron receives each time it comes around the ring. We thus have a situation in which energy is being added to the electrons by the machine and being lost by radiation.

Because they are rather light, electrons radiate much more than heavier particles, such as protons. You can see a bluish glow in electron machines, which results from this so-called synchrotron radiation. Thus, for electrons, the limit at which as much energy is lost through radiation as is added by the acceleration device occurs at a fairly low energy. The largest circular electron accelerators are typically rated at around 7–10 GeV.

To get past this limit, use is made of devices in which electrons are accelerated in straight lines. These are called linear accelerators, or linacs. They, too, were developed in the 1930s and have played an important role in particle physics. The cross section of a typical linear accelerator is shown in Illustration 27. It is a long hollow tube divided at intervals by rings that form separate compartments. Both the tube and the rings are made of a conducting material, such as copper. Each compartment has an independent power supply that can create an electric field. The power supplies are operated in such a way that the resulting electromagnetic wave appears to travel down the tube from one compartment to the next. The electrons "ride" this wave in much the same way that a surfer rides a water wave. As the electrons speed up, the velocity of the wave increases accordingly, so that they stay at the point on the wave where they get maximum acceleration.

The center for research on linear accelerators is at Stanford University, near San Francisco. Starting just after World War II, a series of

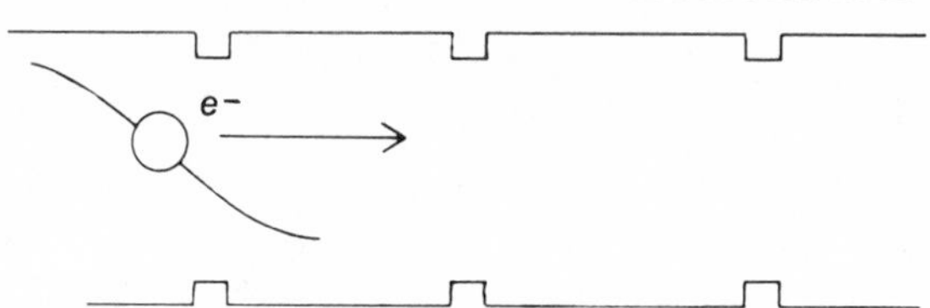

27. Cross section of a typical linear accelerator.

linear electron accelerators (named, appropriately enough, the Mark I, Mark II, and Mark III) eventually produced electron beams of 1.2 GeV. Robert Hofstadter used these machines for detailed studies of the shapes of nuclei and of the proton, for which he was awarded the Nobel Prize in 1961. On May 21, 1966, the first electron beam was brought through the ultimate electron machine—a 2-mile-long accelerator operated by the Stanford Linear Accelerator Center (SLAC). This machine produces 20 GeV electrons and has been the source of electrons used in several of the important discoveries we will describe later.

Secondary Beams and Storage Rings

THE acceleration of electrons and protons remains the center of concern in modern high-energy machines, but over the years a number of other extremely important and useful functions have been developed for these machines. The most interesting of these functions is the secondary beams.

Suppose that the high-energy proton beam from an accelerator is allowed to strike a target as shown in Illustration 28. The target could, in principle, be any material, but it is usually a block of metal, such as copper. When the protons collide with the nuclei of the target, all kinds of particles are produced. These secondary particles emerge from the target in a narrow cone. By running the particles through a suitable arrangement of magnets and slits, we can arrange it so that only the positive pi-mesons (for example) of a certain desired energy come out into the experimental area.* By this process, we can use the primary proton beam to produce a secondary beam of pions and use these pions in our experiments. In this way, it is possible to carry out detailed studies of the interactions of these secondary particles with matter.

*This is known as magnetic selection, a technique to be discussed later in more detail.

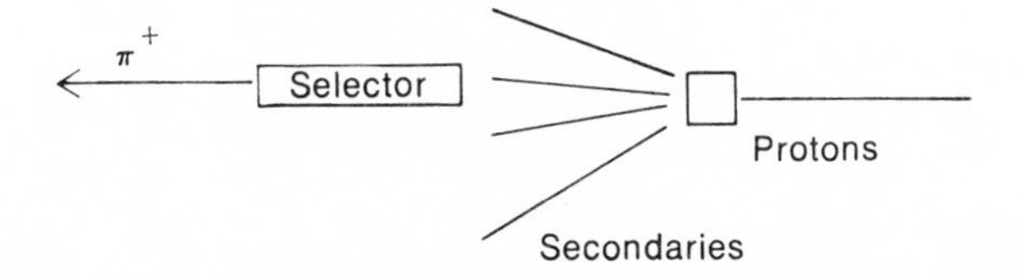

28. The process by which a secondary beam (in this case positive pi-mesons) can be produced.

In modern accelerators, beams of pi- and K-mesons, neutrons, high-energy photons, muons, and neutrinos are routinely available, along with beams of other particles that we have not yet mentioned, such as antiprotons.

If we start with a beam of pi-mesons and wait for a while, they will start to decay into mu-mesons. The mixed beam of pions and muons that results can be "cleaned up" by running it through an appropriate magnet, so that a beam of pure mu-mesons results. This beam is then ready for use in experiments. Over the last decade, as accelerator technology has improved, it has become possible to go one step further. If the mu-meson beam is allowed to go on for a while, the mu-mesons will decay into electrons and neutrinos. If *this* mixed beam is run through a large block of material (typically, steel plates or hundreds of yards of dirt), all the particles will be removed from the beam by interactions, leaving only the neutrinos. The result is a beam of neutrinos. As we shall see later, this beam at the Fermilab has been used to make some important discoveries about the nature of weak interactions.

One more point of interest about secondary beams: The lifetime of the pi-meson is 2.5×10^{-8} second, so if it were traveling at the speed of light you might expect it to go a few centimeters or so before it decayed. Since secondary beams must be meters (and even hundreds of meters) long, do we not face a fundamental contradiction here?

Actually, there is a contradiction unless we apply one of the corollaries of the theory of relativity. The theory tells us that a moving clock appears to run slower than a stationary one (or, more precisely, that a moving clock will appear to be running slower than a stationary one to an observer who is also stationary). The relation between the times measured by the two clocks

$$T_s = \frac{T_v}{\sqrt{1 - v^2/c^2}}$$

where T_s is the time on the stationary clock, T_v the time the moving observer sees on the moving clock, and v the velocity of the moving clock.

If we imagine an observer sitting on a meson, then as far as he is concerned the velocity of the meson is zero, and the meson will decay in 2.5×10^{-8} second, as expected. But if the meson is moving at a high velocity with respect to the laboratory, someone in the laboratory will see the clock that moves along with the meson going much more slowly than his own. Thus, the meson will travel farther than the expected few centimeters in the laboratory.

For example, if the meson is moving at 99.999 percent of the speed of light, then an interval of 2.5×10^{-8} second on the clock moving with the meson translates into

$$T_s = \frac{2.5 \times 10^{-8}}{\sqrt{1 - (0.99999)^2}} = 5.5 \times 10^{-6} \text{ sec}$$

for a clock in the laboratory. In this time, the meson will travel a distance $D \approx 3 \times 10^8 \times 5.1 \times 10^{-6} = 1{,}530$ meters in the laboratory —ample space for the construction of the secondary beam. Hence, the fact that secondary beams exist at all can be taken as evidence for the theory of relativity!

A second important accessory to the modern accelerator is the storage ring. This device is a set of magnets in a ring, a design much like the main ring of a synchrotron, but without accelerating devices. Pulses of particles are fed into the ring from an accelerator and the magnets are adjusted so that the pulse of particles will keep circulating. In this sense, the accelerated particles are stored in the ring.

A typical storage ring situation looks like Illustration 29. Particles from an accelerator are fed into two rings over a period of time until the rings are "filled." This operation might take 30 minutes. Then particles from the two rings are brought into an interaction area where they are allowed to collide head on. Just as a head-on car crash releases more energy than a crash with a stationary object, head-on collisions in storage rings provide more energy with which to make new particles.

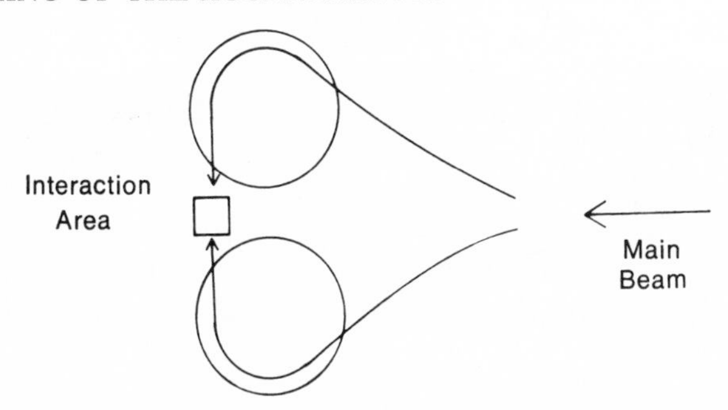

29. A typical storage ring structure.

There are storage rings for protons, electrons, and positrons. Various combinations (e.g., electrons in one ring and positrons or protons in the other) are either in existence or in advanced planning stages. We will discuss some important storage ring experiments in Chapter XII.

Fermi National Accelerator Laboratory: A Typical (but Large) Machine

JUST to get some idea of what a big accelerator actually looks like, let us talk about the Fermi National Accelerator Laboratory. Located about 50 miles west of Chicago, it is the world's largest accelerator. See Illustration 30, below, for a schematic of the machine.

The protons to be accelerated are first created by ionizing hydrogen. They start their journey by being boosted to 750 keV by an ordinary high-voltage device, and are then put into a 200-MeV linear accelerator. From this device they are fed into a "booster," which is actually an 8-GeV synchrotron. Once they have achieved this energy, they are injected into the main ring. You can think of the three preaccelerators as playing roles somewhat analogous to the gears in an automobile transmission. Each one gets the proton going faster until we are ready to shift into "high."

In the FNAL, "high gear" corresponds to a ring 1 kilometer in radius. About thirteen pulses from the booster are required to fill the ring, a process that takes about 1 second. Once the ring is filled, the acceleration process starts. Kept in the ring by 1,000 magnets, the protons make about 70,000 revolutions, picking up 2.8 MeV per turn from the sixteen

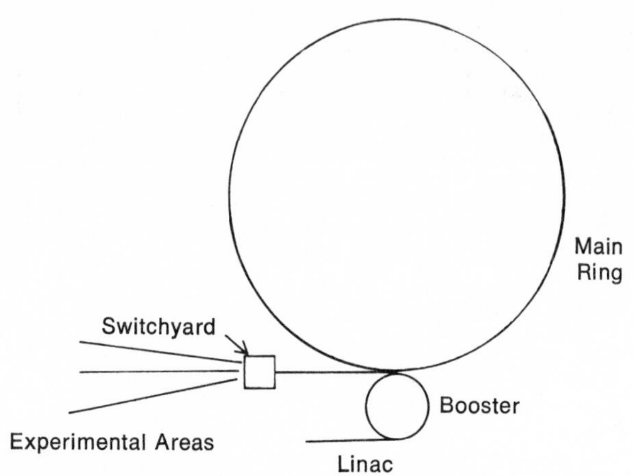

30. Sketch of the Fermilab accelerator.

accelerating stations around the ring. Finally, they emerge (after about 3 seconds) as 200-GeV protons. For 500 GeV, it takes longer.

Once the protons are extracted from the beam, they are directed (in an area called the beam switchyard) to one of three experimental halls. These are the proton area, where the protons themselves are used for experiments, and the meson and neutrino areas. The meson beams are made by the method described in the previous section. Because of the high speed of the particles, the actual experimental setups are located over 300 meters downstream from the point at which the secondary particles originate. It takes that much space to get the various different mesons (and neutrons) sorted out.

The mesons that ultimately give rise to the neutrino beam travel about 300 meters before encountering an earth barrier 1 kilometer long. The neutrinos that emerge go to an experimental area that is fully 2 kilometers from the main ring. When you deal with high energies, you clearly have to start thinking on a big scale.

In the 40 years since Lawrence's first 4-inch cyclotron, accelerators have come a long way, both in size and in the amount of knowledge they have given us about the world. Most of the rest of this book will be devoted to describing and trying to understand things that have been found in accelerator experiments.

CHAPTER VII

The Proliferation of Elementary Particles

Things were not slow in becoming curious.

—THOMAS PYNCHON,
The Crying of Lot 49

The Discovery of the Antiproton: A Typical Accelerator Experiment

BY the early 1950s, a number of lines of research came together in a way that made the next big push in elementary particle work possible. The cosmic ray data had shown that there were many more particles than had been expected. One textbook went so far as to title a chapter "Particles We Might Do Without." At the same time, it was pretty clear that cosmic ray experiments had reached the limit of the results they could produce, and the advent of accelerators in the GeV range came at just the right time to keep things going. One of the major experiments that was performed resulted in the discovery, at Berkeley, of the antiproton.

We have already encountered the idea of antiparticles in connection with the positron. That a particle should exist with the same mass as the proton but with a negative electrical charge was accepted on faith by most theorists. There were even a few cosmic ray events that could, with some stretch of the imagination, be interpreted as evidence for such a particle. Getting hard laboratory evidence, however, was another matter. With a proton beam, the only way an antiproton could be produced was through the reaction $pp \rightarrow ppp\bar{p}$. The symbol $\bar{p}$ is to represent the antiparticle. (We will use the convention of the bar over the particle symbol to denote antiparticles from this point on.) The reason that the production of an antiproton has to be accompanied by the production of an extra proton arises from conservation laws that we will discuss later.

How much energy is needed to create the antiparticle? The first impulse is to say that since two extra particles of mass 938 MeV have to be produced, we need a beam with kinetic energy $2 \times 938 = 1{,}876$ MeV. It turns out that energy and momentum conservation require that the four particles in the final state of the reaction cannot be sitting still: They must have some minimum amount of kinetic energy. When this is taken into account, the kinetic energy of the proton from the accelerator has to be about 5.6 GeV. This requirement was very much in the minds of the men who designed the bevatron at Berkeley. In fact, that machine has been characterized as the accelerator "designed to produce the antiproton."

Once the required energy was achieved, the problem that faced the experimenters was how to tell which of the many negative particles produced in collisions were antiprotons, and, most important, how to distinguish them from negative pi-mesons. There were several tools available, which should be described separately before we can grasp how they were put together in an experiment.

We know that when particles of a given mass and velocity move into a magnetic field, they move in a circle whose radius is given by

$$R = \frac{mv}{Bq} = \frac{P}{Bq}$$

where we have replaced the quantity mv in the second equality by the letter P. This quantity is known as the momentum of the particle.

From this equation it follows that if two particles of different momentum enter a magnetic field, each will start to move in a circle with a different radius. For example, in Illustration 31 we show two particles of momentum P_1 and P_2. The orbits of these particles will be as shown, where the radii R_1 and R_2 are given by the equation above.

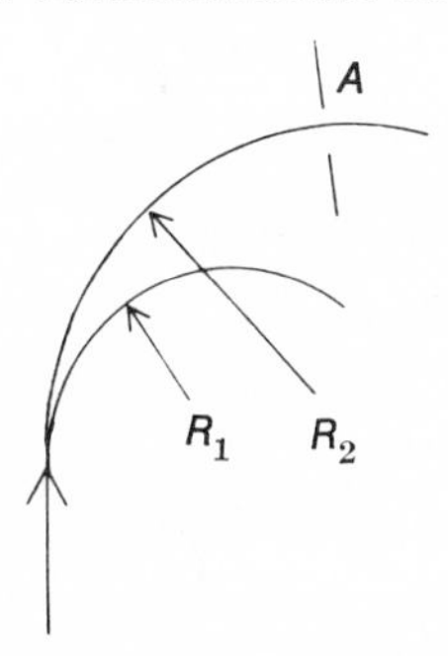

31. The magnetic spectrometer.

Suppose we put a narrow slit at point *A*, as shown. A particle of momentum P_1 will pass through the slit, but a particle of momentum P_2 (or any momentum other than P_1) will hit the solid material. As a result, on the other side of the slit we would see only particles of momentum P_1. In this way, we can say that the magnetic-field-plus-slit arrangement "selects" momentum P_1 and rejects all others. This apparatus is called a magnetic spectrometer, or momentum analyzer.

Imagine, then, that particles produced in a collision are run into such a magnet. If we are looking for negative charges, then positively charged particles will curve in the opposite direction to the orbits shown in this illustration, and will come nowhere near the slit. Furthermore, only those negatively charged particles with a preselected momentum will pass through the slit. Thus, this simple operation will immediately reduce the unwanted particles by a large factor.

However, it is possible for a negative pion and an antiproton to have the same momentum even if they have different masses. All that is required is that their mv be the same. Another technique must be used that will give an independent determination of the particle velocity after the momentum selection has been made. The simplest way to do this is shown in Illustration 32. The beam of particles is passed through two thin layers of scintillating material a known distance D apart. A particle will cause a flash of light from each of these as it passes. By measuring the time between these flashes, the velocity of the particle can be determined. This is called a time-of-flight measurement, and is quite common in experimental physics. Obviously, its success depends critically on the experimenter's ability to build fast, accurate electronic timers.

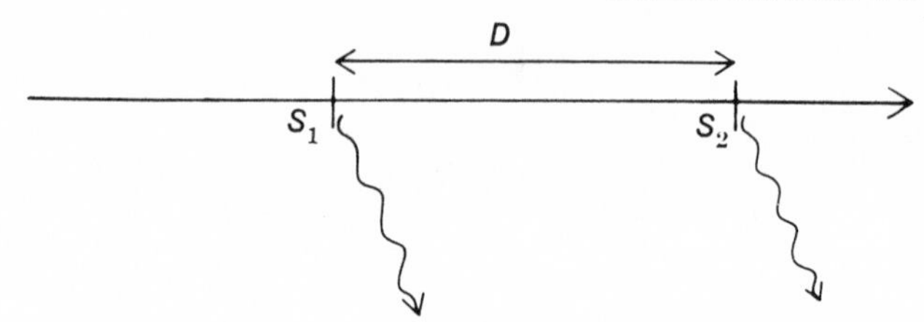

32. A time of flight measurement. Signals from S_1 and S_2 indicate how long it took the particle to travel a distance D.

A second way of determining the velocity of fast particles was developed by the Russian physicist Pavel A. Čerenkov in 1934. When a fast particle enters a material such as a gas or liquid, it can emit a burst of light. This light, called Čerenkov radiation, is somewhat analogous to the sonic boom emitted by supersonic aircraft. From our point of view, the important aspect of this radiation is that the angle at which it is emitted depends on the velocity of the particle. A typical Čerenkov counter is sketched in Illustration 33. From the angle θ we can determine how fast the particle is moving through the medium.

With this background in experimental technique, we are in a position to understand the details of the antiproton search. The apparatus is sketched in Illustration 34, on page 101. The proton beam from the accelerator strikes the target, thereby creating a flood of particles and, perhaps, an antiproton or two. The secondary particles are run through a magnet to select the proper momentum, and are then brought out through the concrete shielding to a scintillation counter S_1, another magnet, and a second scintillation counter S_2. From time-of-flight measurements, the velocity of the particle is now known. The beam contin-

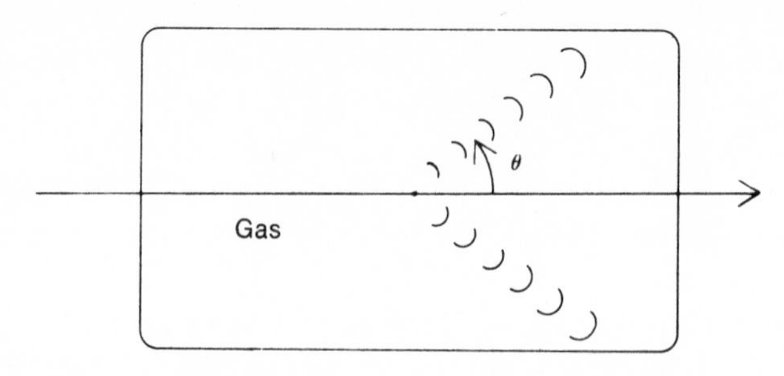

33. A typical Čerenkov counter.

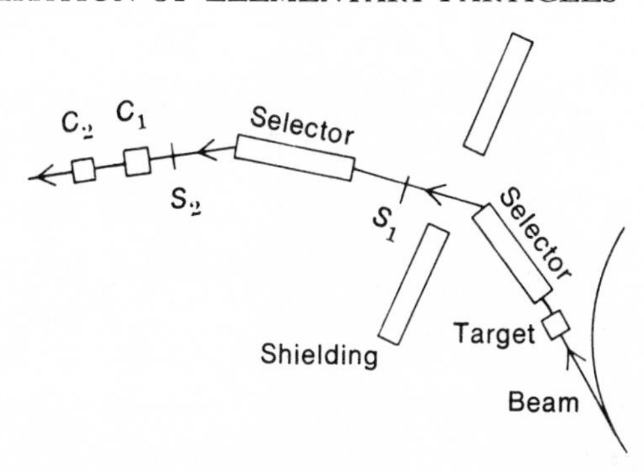

34. Apparatus used in the antiproton search.

ues to two Čerenkov counters, the first of which is set to register if a particle faster than an antiproton goes through it, and the second of which is set to register if a particle having the correct speed for an antiproton is there.

Why are the Čerenkov counters there at all? Is the time of flight not a sufficient criterion for distinguishing between antiprotons and mesons?

In principle, one velocity measurement should suffice to identify the antiproton. But in an experiment of this type it is not at all unusual for the beam coming from the selector to contain tens of thousands of pions for each antiproton. It might be possible, therefore, for one pion to go through the first scintillator and for a different pion to go through the second in just such a way as to produce a spurious time of flight that is exactly equal to that of the antiproton. Such "accidental coincidences" are the bane of the experimental physicist's life. The Čerenkov counters are one way of dealing with them.

To identify a particle as an antiproton, three requirements must be met: The time of flight must be correct; the second Čerenkov counter must register; and the first Čerenkov counter must not register. (In technical terms, we say that we "veto" with this counter.) Only if all three of these conditions are satisfied can we be sure we have seen the particle we are looking for and not some accidental collection of pions.

This sort of careful "overkill" on the identification of particles is typical of accelerator experiments. In the antiproton experiment a total of fifty particles were seen in months of work, so we could think of the search as an analog of the needle in the haystack. For their discovery of the antiproton, Owen Chamberlain and Emilio Segrè of Berkeley were awarded a Nobel Prize in 1959.

The Pion-Nucleon Resonance

IN Chapter VI we saw how the intrinsic limits of the cyclotron made it impossible to produce pi-mesons. Nonetheless, the artificial production of pions remained an important goal in physics. They are, after all, the lightest of the strongly interacting particles, and therefore the easiest to produce. In addition, they are the primary particle involved in the nuclear force. Therefore, whether the purpose is to study the pions themselves or to use them to study other interactions, producing them becomes a very important goal.

In the late 1940s and early 1950s, before the synchrotrons became widely available, a machine called the synchrocyclotron was used for this purpose. This machine was a cross between the cyclotron and the synchrotron: Like the former, it guided particles with a constant magnetic field; like the latter, it accelerated particles in bunches. It overcame the cyclotron limit by lowering the rate at which the voltage across the gaps was changed as the particles moved toward larger radii. In this way, the slowing down of the particles because of increased mass could be taken into account. In a typical machine (like the one installed at the University of Chicago), the diameter of the magnets might be 170 inches (almost 15 feet) and the frequency might drop by 60 percent as the particles spiraled out. With such a machine, proton energies in the range of 400–500 MeV could be attained—enough to produce pi-mesons.

One of the first things measured when pion beams became available was the way in which pions interacted with protons. This information could be obtained by letting the pion beam strike a target composed of hydrogen and seeing what happened when the pion hit the hydrogen nucleus. One of the simplest questions to ask was this: "If a pi-meson comes near a proton, what are the chances that the two will interact with each other?"

In the diagram (Illus. 35) we show two pions in a beam coming near a proton target. Pion *A* is not deflected at all, so we would say that it did not interact with the proton. From an experimenter's point of view,

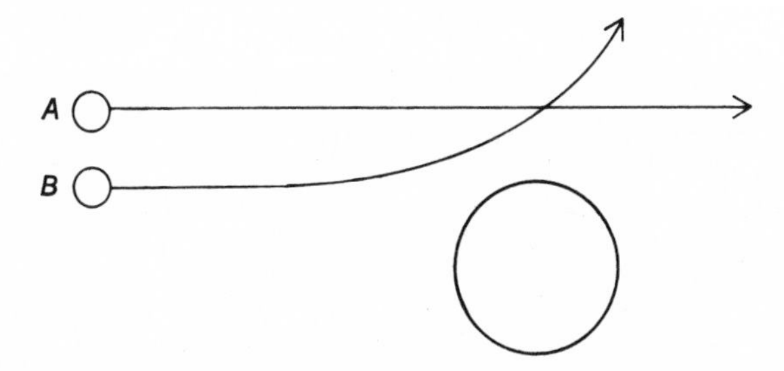

35. Pion *A* does not interact with the proton, while pion *B* does.

we could recognize this situation by noting that pion *A* remained in the beam after it had passed the target. Pion *B*, on the other hand, does interact with the proton. This interaction could be through the electrical force if the pion is charged, or it could be through the strong interaction. In either case, the result is the same. The pion is scattered out of the beam.

Physicists usually express interaction probabilities in terms of something called a *cross section.* If you imagine holding a circular disk in the pion beam in Illustration 35, it is clear that the disk will also scatter mesons out of the beam. The cross-sectional area of the disk that will scatter out just as many pions as does the proton is called the *pion-proton cross section.* It can be measured by allowing a beam with a known number of pions to enter a target that has a known number of protons in it. By counting the number of pions left in the beam after it has traversed the target, we can deduce the interaction probability, and, hence, the cross section.

Starting in 1952 with a group under the direction of Enrico Fermi at Chicago, physicists began collecting data on the scattering of charged pions from hydrogen. When they plotted their results, a graph similar to the one shown in Illustration 36 emerged. There was a large peak in the cross section at a pion energy of about 200 MeV—a peak that was about 100 MeV wide.

This sort of bump in a cross section means that when the pion has exactly the right energy with respect to the proton, it is much more likely to interact than if it has some other energy. One way of thinking about these situations is to imagine that at this precise energy, the pion and the proton can "lock together" for a short time, whereas at other energies they just bounce off one another. If we think of things this way,

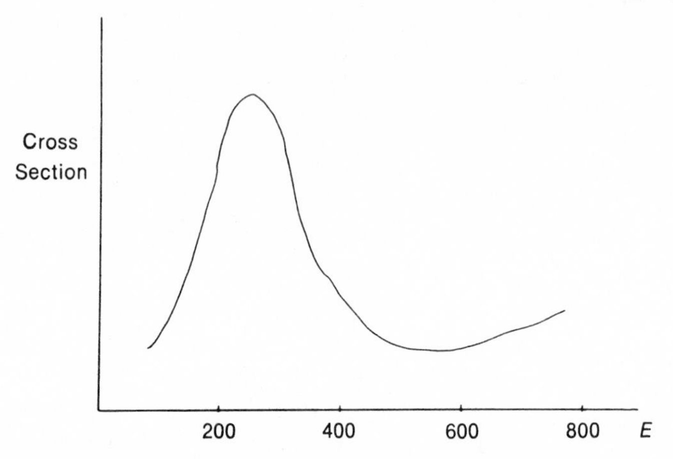

36. The cross section for the scattering of a positive pi-meson from a proton, showing the delta peak.

there is no reason why we should not think of the locked-together state as a particle. If we do so, we can represent the interaction of a pion and a proton near the peak of the cross section with a diagram similar to the one below (Illus. 37). The two particles come together and fuse into another particle which, after a short while, separates into the original pion and proton again. The intermediate particle in this diagram is called a *resonance,* and in modern terminology is denoted by the Greek letter Δ (delta). Since the particle in the diagram must have two positive charges, it is written Δ^{++}.

How long would such a particle live? One way of estimating the lifetime is to use the uncertainty principle we introduced in Chapter III. We can see from Illustration 37 that the Δ must have an energy equal to the sum of the pion and proton energies. But what is the uncertainty in the energies required to produce the particle? Clearly,

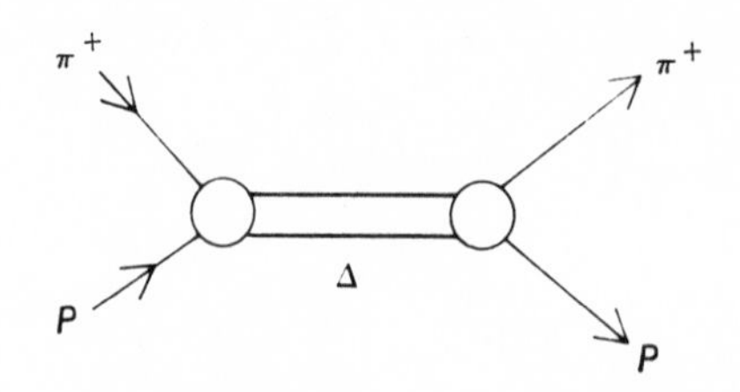

37. Interaction of a pion and a proton near the cross section's peak.

if the pion is at the energy corresponding to the peak of the bump in the cross section, the resonance will be formed. But what if the pion energy is a little lower or higher, so that it corresponds to a point one-third of the way down the peak? Or one-half or one-fifth? Will these energies correspond to Δ^{++} production as well? In fact, if we follow this principle, it seems reasonable to say that ΔE, the uncertainty in the energy of the resonance, is roughly the width of the peak in the cross section. For the Δ^{++} this is about $\Delta E \approx 100$ MeV $= 1.6 \times 10^{-4}$ erg, so from the uncertainty principle, the uncertainty in the time must be

$$\Delta t = \frac{h}{\Delta E} = \frac{6.6 \times 10^{-27}}{1.6 \times 10^{-4}} \sim 4 \times 10^{-23} \text{ sec}$$

This, in turn, can be taken as a reasonable estimate of the lifetime of the Δ^{++}.

This lifetime is very short compared to the life of any particles discussed so far. In fact, physicists in the 1950s were very reluctant to consider resonances as particles at all. One author went so far as to compare them to automobiles that fell apart before they left the factory. The reason for this attitude rests, I think, on the fact that it is so difficult to detect particles with short lifetimes. The strange particles live long enough to travel several inches in a cloud chamber, so their tracks can be seen in the usual way just by looking at the droplets. It requires a stretch of the imagination to extend the term *particle* to something that can be seen only indirectly and that never leaves a visible track in a detection device.

But the idea that something is a particle only if it is easy to detect seems a little artificial. The delta, after all, lives long enough to travel from one side of a nucleus to the other; therefore, its lifetime is quite respectable compared to the characteristic time of the strong interactions. Because many particles with lifetimes comparable to the delta have been discovered, and because these particles seem to play an important role in the strong interactions, physicists have become accustomed to applying the term particle to them.

To complete the story on the delta, careful study of the different pion-nucleon scattering cross sections shows that they are a family of particles, similar to the pions. They come in four charge states: Δ^{++}, Δ^{+}, Δ°, and Δ^{-}, where the superscripts refer to two positive charges, one positive charge, neutral, and one negative charge, respectively. The mass of the family is about 1,236 MeV.

Meson Resonances

THE idea that a bump in a scattering cross section can be interpreted as evidence for a short-lived particle immediately suggests that there might be resonances in systems other than those involving the pion and the nucleon. We could ask, for example, whether there might be a resonance in the cross section for the scattering of one pion from another, or in the scattering of a pion from a Λ°. Unfortunately, although it is possible to make beams of pi-mesons, none of the other particles we have studied live long enough to be made into targets. Consequently, the direct identification of resonances in such systems cannot be accomplished in the same way as it was for the delta.

If you think of the mechanism by which a resonance is formed, however, you will realize that it is not really necessary to have a conventional beam-plus-target arrangement to make one. All that is necessary is that the particles that are to form it be near each other for a period of time that is characteristic of the strong interaction. This can happen in a target-beam experiment, of course, but it can also happen when the two resonating particles are produced in the same reaction. For example, it is possible to start with a beam of pions and have a reaction such as $\pi^- P \rightarrow \pi^+ \pi^- n$, in which two pions are present in the final state. These two pions will interact with each other just as surely as they would if one were a target and the other a beam. Consequently, we can ask about whether they interact in such a way as to form a resonance or not.

If the particles form a resonance, then the reaction described above can be represented schematically as in Illustration 38. Here, again, the double line represents the resonance. In order to see whether the data support this kind of picture, we can look for bumps again—not bumps in a cross section, as in the case of the delta, but bumps in something called a phase space diagram.

We start this procedure by imagining that we are riding on the resonance in the previous diagram. After the resonance decays we will see the two pions moving off. Each of them will have some kinetic energy that we could, in principle, measure. We could then make a plot of the

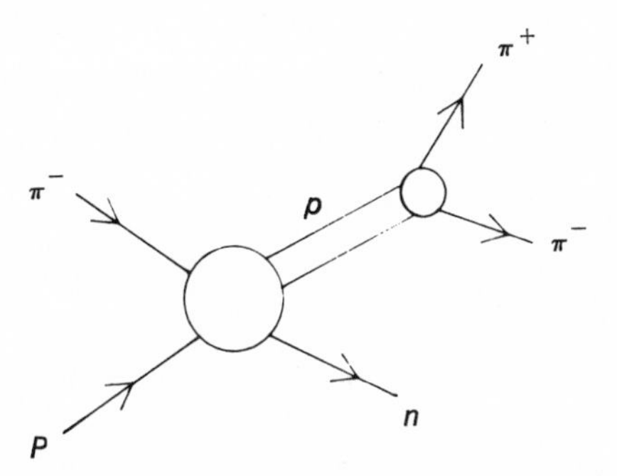

38. The process by which a rho-meson is produced and decays.

number of pion pairs produced with a certain total energy as a function of that energy. The result of such a plot could be either of the graphs in Illustration 39. In the left-hand graph the pion pairs are distributed evenly over the allowed range of energies. Such a graph would be interpreted as evidence that the two pions were produced independently of each other and therefore as evidence for the absence of a resonance. The right-hand graph, however, shows an excess of pion pairs at a given energy. This bump is what corresponds to the peak in a cross section. It tells us that there is some sort of interaction between the pions that causes them to be produced preferentially at this particular energy, which, in turn, means that there must be a resonance between the two particles.

The idea behind this method of analysis is that *if* we could make a pi-meson target and bombard it with a pi-meson beam, we would see a bump in the cross section corresponding to a graph similar to the one

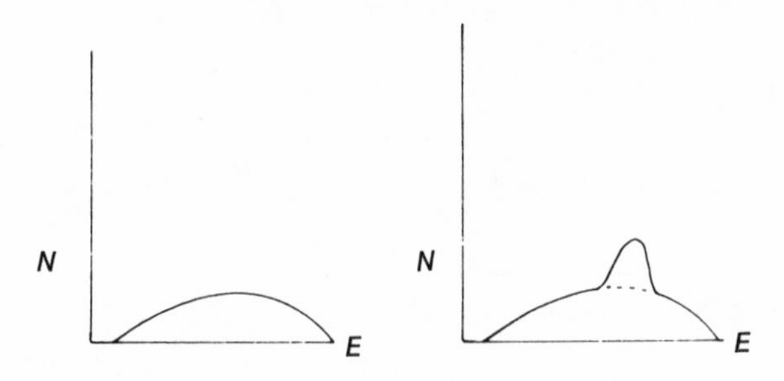

39. The number of particle pairs of a given energy for no resonance *(left)* and with a resonance *(right)*.

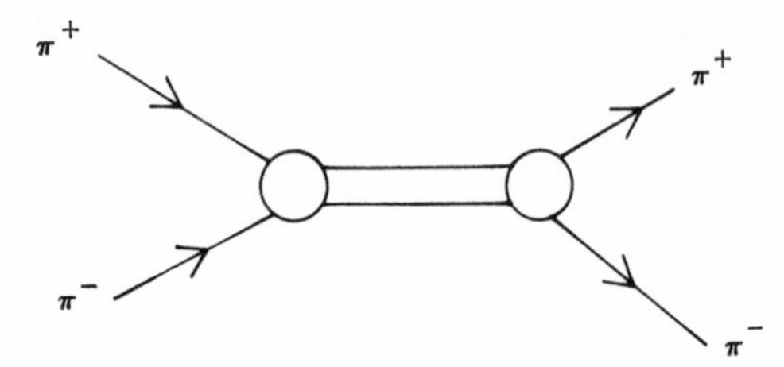

40. The production of a rho in a mythical pion–pion scattering experiment.

in Illustration 40. Just as a bump in the cross section led us to the delta, a bump in the phase space diagram leads us to other resonances.

In 1961 a group of scientists working at Brookhaven carried out the same kind of analysis that we have been describing and got results similar to those pictured in Illustration 41. There was a definite peak at around 760 MeV, with a width of a little over 100 MeV. This new particle lives long enough to participate in the strong interactions, but not long enough to leave a track in a cloud chamber, which is also true of the delta. The new particle is called the ρ (rho) meson, and like the pion, it comes in three charge varieties—that is, positive, neutral, and negative.

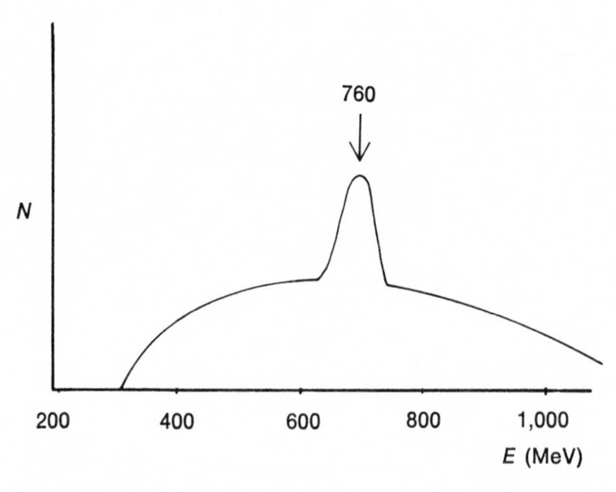

41. The phase space diagram, which shows the existence of the rho-meson.

Life with the Strange Set

THE 1950s also proved a productive time in the realm of strange particles. You will recall that these were particles that decayed rather slowly so that they were visible in cloud chambers, either directly or through their decay products, as a V particle. About the same time that the Brookhaven accelerator was coming on-line, the last in a series of cosmic ray discoveries were made. Not one, but two, new strange particles were discovered, each heavier than a proton.

In 1953 a particle of mass of about 1,190 MeV was seen. It was denoted by the Greek letter Σ (sigma). We now know that there are, in fact, three particles in the Σ "family," one with positive charge, one with negative, and one neutral. The charged particles decay into a nucleon and a pion in about 10^{-10} second, while the neutral Σ decays somewhat more quickly into a lambda and a photon. Like the lambda, this particle was totally unexpected and seemed to play no known role in physics.

Then, in 1954, the last of the "lucky" cosmic ray events was recorded. This time it was a particle of mass about 1,320 MeV, which decayed in 10^{-10} second into a lambda and a pion. Because the lambda then decayed into a nucleon and a pion, this particle was seen as the start of a "cascade" of decays, and was christened the *cascade* particle. Following the custom of giving particles letters of the Greek alphabet, this particle was denoted by the letter Ξ (xi). There are two xi particles, one neutral and one with a negative charge.

One thing to note about the cascade is that not only does it decay slowly itself, but its decay products include a particle, the lambda, which also decays slowly. Therefore, whatever it is that makes the lambda and the sigma "strange," should appear in a double helping in the cascade. We might even label it *doubly strange.*

The appearance of all these strange particles set the theoretical community to wondering, and by 1957 some interesting ideas started floating around. If the nucleon has a resonance (the delta), why should the strange particles not have resonances as well? In 1961 a group at Berkeley looked at the reaction $k^-p \rightarrow \Lambda^\circ\pi^+\pi^-$ and did a phase space analysis of the lambda-pion state. Sure enough, a bump showed up with a mass of about 1,385 MeV and a width of about 40 MeV. This particular strange resonance decayed into a lambda and a pion in about 10^{-22} second, the characteristic time scale for strong interactions. From this we conclude that not every decay involving a strange particle has to be slow. This is a puzzling fact, but it is a fact. This particular strange resonance is now called the $\Sigma(1385)$. (Nomenclature will be discussed in Chapter IX.)

The Bubble Chamber

EXPERIMENTS of the type used to discover the antiproton are known generically as *counter experiment.* Provided that we have a pretty good idea of what we are searching for, they provide a very precise way of measuring particle properties. In the cosmic ray experiments, however, we saw that some of the most important results are those that are not expected. The ability to visualize the particle events provided by the cloud chamber was very important in the early work in particle physics.

In order for an interaction to be seen in the chamber, however, it is necessary that the incoming particle collide with a nucleus while it is passing through. When the energies of the incoming particles get into the GeV range, such a collision becomes less and less likely. For a cloud chamber to be effective in a modern accelerator beam, for example, it might have to be 100 yards across. A shorter chamber would simply not provide enough interactions to study the incoming particles. Clearly, construction of a cloud chamber on these dimensions is out of the question.

Consequently, visualization for high-energy particles is accomplished by a device somewhat similar to the cloud chamber, but without its disadvantages. It is called the bubble chamber.

Like the cloud chamber, the bubble chamber depends on the ionization of atoms in the wake of a fast particle in order to provide the centers around which the track will develop. Unlike the cloud chamber, the bubble chamber is filled with liquid held near its boiling point and under pressure. Instead of using the ions as centers of condensation, the ions serve as centers around which boiling occurs in the fluid. You can see a very similar phenomenon when you open a bottle of carbonated beverage. Bubbles start to rise to the top of the bottle, but if you look closely you will notice that the bubbles come in streams from definite spots on the inner surface of the bottle. These are spots where some local roughness provides a nucleus around which the bubbles can form.

The folklore in the physics community has it that Donald A. Glaser, the man who developed the bubble chamber at the University of Michigan and who received the Nobel Prize in 1960, got his inspiration when he opened a bottle of beer in an Ann Arbor saloon. (The folklore does not include the brand name).

The bubble chamber operates as follows: A particle passes through the fluid (which is ready to boil if a nucleus for bubbles appears), leaving a trail of ions in its wake. A piston on top of the chamber is then pulled out quickly, lowering the pressure on the fluid, which starts to form bubbles around the ions. The bubbles are photographed, the ions are "swept" out of the chamber by a magnetic field, the piston is lowered, and the whole cycle can be repeated. Typically, the chamber will be operated each time a pulse of particles from an accelerator comes through, and the resetting goes on while the next pulse is being accelerated.

The size of a bubble chamber is normally measured in meters. Modern low-temperature technology has led to the construction of bubble chambers in which the working fluid is liquified hydrogen. In chambers of this type, the particles in the incident beam interact with hydrogen nuclei (i.e., protons), and these interactions are photographed as described above. In this way, the fluid serves both as the experimental target and the means by which the charged particles are detected. In such chambers, it is possible to run experiments of the "Let's put the beam in and see what happens" variety, and also to look for very rare situations where only a handful of events of the desired type might be expected. In addition, using bubble chambers makes it possible to record data on all the final states in a given reaction, so that the search for resonances can be carried out in each of the states separately after the experiment is finished. For example, in an experiment where a pion beam enters a hydrogen bubble chamber we could study the delta by looking at the reactions where only a pion and a nucleon are produced in the final state, and the rho-meson by looking at other reactions in which two pions (and a nucleon) are in the final state.

The Proliferation Orgy

IF you are confused by this time, imagine how physicists must have felt in the early 1960s. It seemed that every time they turned around someone was discovering another particle. Far from solving the problem of the strong interactions, the advent of high-energy accelerators and sophisticated detection equipment seemed to lead to more and more confusion. As accelerators in the 20–30 GeV range came on-line at Brookhaven and at the European Center for Nuclear Research (CERN) in Geneva, the discovery of new particles (primarily of the resonance variety) became an everyday occurrence. In fact, bump hunting became something of a cottage industry for physicists throughout the 1960s.

Perhaps a nontechnical example will serve to illustrate this point. In 1963, the Finnish physicist Matt Roos compiled the first comprehensive table of elementary particles and resonances. Published in the *Review of Modern Physics,* Roos's article contained two tables and ran about five pages. There were seventeen "particles" and twenty-four "resonances." This article appeared periodically after that, expanding and being updated each time it appeared. In 1972 the distinction between particles and resonances was finally dropped. The last edition of the article, in 1976, ran 245 pages (with a thirty-page supplement) and took up an entire edition of the journal. Literally hundreds of particles are now known and cataloged in these tables.

Obviously, when particles start appearing in such profusion, the search for order among them takes on a very high priority. This will be the topic of the Chapter VIII. Before we turn to it, there is one question that we ought to face.

We started the search for elementary particles in the hope of finding simplicity in nature. For a while it seemed we had succeeded, but the developments we have just described appeared to say otherwise. If there are more elementary particles than there are chemical elements, then the world has not been made any more simple than it was before. Obviously, the particles that have been discovered cannot be "elementary" in the intended sense of the word.

There is, in fact, a strong temptation to say that most of the very short-lived particles are not really elementary at all. We have already seen how reluctant physicists were to apply the label particle to them. But if we are going to exclude some of the newly discovered particles from the class elementary, we must have some criterion for doing so.

What will it be?

Stability? If we excluded all unstable particles, we would have to throw out the neutron and the pion, both of which are necessary to explain nuclear structure.

Lifetime? If we kept only long-lived particles, we would still have the puzzle of strangeness. In addition, we know that resonances like the rho and delta play a much more important role in keeping the nucleus together than do relatively obscure but longer lived particles, such as the cascade. Finally, saying that a lifetime is "long" or "short" implies that there is some standard of comparison against which a lifetime can be judged. For particles, the only reasonable standard is the characteristic time of the strong interaction. By this standard, *every* particle ever discovered is "long-lived," since all of them have lifetimes that would allow them to participate in the interaction.

Hence, there is no logically consistent way of calling some particles elementary and other, nonelementary. This situation was called nuclear democracy by Geoffrey F. Chew of Berkeley. Basically, as far as the strong interactions are concerned, all particles are equally important. We simply have to accept the fact that there are a lot of them and see where it leads us.

CHAPTER VIII

The Search for Order amid Chaos

A place for everything and
everything in its place.

—ANONYMOUS

Introduction

THERE are two different senses in which one can talk of imposing order on a large group of seemingly unrelated objects, such as the elementary particles. For the sake of discussion, let us call them classification and reduction. The difference between the two can be illustrated by an architectural example.

The buildings in a large city would seem to form a class of unordered objects. No two are exactly the same, and a school might be found next to an apartment building, a high-rise office block next to a residential area, and so forth. If we were to impose an order on these buildings by classification, we would start by trying to find groups of buildings with common characteristics. When we had found such a group, we would give it a name and then lump under the one name all the buildings that shared this characteristic. In this way, the large number of individual buildings would be replaced by a few categories of types of buildings, and most people would agree that the situation was more orderly.

The classification of buildings could also be based on use—residential, commercial, and industrial (these are, in fact, the categories used by the Census Bureau). We could classify them by type of construction—wood frame versus steel and concrete. We could classify them by height, by year of construction, by value, or by any other criterion that we considered useful. I can recall being involved in a study of solar-energy potential in a city where we were trying to classify homes by the direction in which the roof pointed.

One important point about building classification schemes such as this is that an individual building may very well belong in several categories. Think of a wood-frame one-story residential building with a roof facing south that was worth $50,000. Which of the different attributes of the house was important would depend on your ultimate purpose. The price would be important if you were buying it, but the orientation of the roof would be more important if you were thinking about installing a solar collector.

If we took the same city and attempted to impose order by reduction, we would proceed in a different way. We would start by taking some buildings apart (either literally or figuratively) to see what they were made of. We would then start putting together a list of the kinds of materials we found—lumber, bricks, roofing shingles, glass, and so on. This list would probably not be long. We would then say that the materials in the list constituted the elements of buildings in the city, and that every structure was made of these elements arranged in different ways. The long and complex list of individual buildings would be replaced by a short list of basic building materials, and this, too, would result in a more ordered way of thinking about the city.

To use an analogy from science, someone who wants to find order among the collection of chemical elements can either group them according to their chemical properties (as Dmitri I. Mendeleev did when he put together the periodic table of the elements) or the researcher can find the basic constituents of the atoms (as Rutherford did in evolving the modern atomic theory). These are simply complementary ways of approaching the same problem.

Both classification and reduction have been used to bring the proliferation of elementary particles under control. In this chapter we will take up the various ways of categorizing these particles that have proved useful, and in Chapter IX we will discuss the idea that the particles we have seen are made up of a small number of basic building blocks.

Classification by Interaction: Leptons and Hadrons

We have so far listed three different types of interactions that can affect an elementary particle. All charged particles are affected by the electromagnetic force, and hence can be said to participate in the electromagnetic interaction. Most of the particles we have studied are either created or decay via the strong interaction, and a few seem to be involved only in the weak interaction. We can use this difference in the type of interaction to introduce one way of classifying particles.

The electron, the muon, and the neutrino do not seem to be part of the strong interaction at all. These three particles are called *leptons* (weakly interacting ones). This is a rather small group of particles, which is important mainly in the study of slow decays and other weak interactions.

All the other particles we have discussed, with the exception of the photon, are involved in one way or another with the strong interactions. They are called *hadrons* (from the Greek root hadrys, or strong). The proliferation of particles that we discussed in Chapter VII is entirely in this category. Consequently, most of the effort that physicists have made to try to sort out and categorize elementary particles has involved the hadrons.

In a scheme where particles are classified by interaction, the photon is generally put in a class by itself, since it is the particle that mediates the electromagnetic interaction.

Classification by Decay Product: Mesons and Baryons

If we watch any hadron long enough, we will eventually see it decay into some collection of the stable particles—the proton, electron, photon, and neutrino. A possible decay scheme for the negative cascade particle is shown to illustrate this point:

$$
\begin{array}{l}
\Xi^- \to \Lambda^\circ + \pi^- \\
\quad\quad\quad \pi^- \to \mu^- + \nu \\
\quad\quad\quad\quad \mu^- \to e^- + \nu + \bar{\nu} \\
\quad \Lambda^\circ \to n + \pi^\circ \\
\quad\quad\quad \pi^\circ \to \gamma + \gamma \\
\quad\quad n \to p + e + \bar{\nu}
\end{array}
$$

Some of these decays will be fast and some will be slow, but the end products are the stable particles.

There are two possible ways in which decay chains of this type can come out. There may be only leptons and photons in the final collection, or, as in the example above, there may be a proton as well. The presence or absence of a proton therefore becomes a criterion that we can use for classification.

Particles, such as the cascade, in which a proton does appear in the end product of the decays, are called *baryons* (heavy ones). The proton itself is included in this class, as are the lambda, the sigma, and the delta.

Particles whose final collection of decay particles is made up entirely of leptons and photons are called *mesons*. This definition of the term meson now supersedes the original one, in which the meson was thought of as a particle intermediate in mass between the proton and electron. The pi- and *K*-mesons obviously satisfy both definitions of the word. For example, the decay scheme for a π^+ is shown below:

$$
\begin{array}{l}
\pi^+ \to \mu^+ + \nu \\
\quad\quad\quad \mu^+ \to e^+ + \nu + \bar{\nu}
\end{array}
$$

On the other hand, with the new definition it becomes possible to talk about mesons that are more massive than the proton. Many particles of this type have been discovered. For example, there is a particle of the resonance variety called the A_2 meson with a mass of 1,310 MeV, which decays by a series of fast and slow decays as follows:

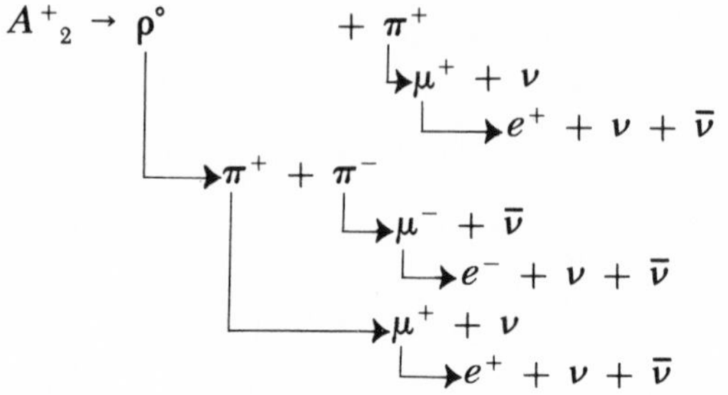

$$
\begin{array}{l}
A^{+}_{2} \rightarrow \rho^{\circ} + \pi^{+} \\
\quad \pi^{+} \rightarrow \mu^{+} + \nu \\
\quad\quad \mu^{+} \rightarrow e^{+} + \nu + \bar{\nu} \\
\quad \rho^{\circ} \rightarrow \pi^{+} + \pi^{-} \\
\quad\quad \pi^{-} \rightarrow \mu^{-} + \bar{\nu} \\
\quad\quad\quad \mu^{-} \rightarrow e^{-} + \nu + \bar{\nu} \\
\quad\quad \pi^{+} \rightarrow \mu^{+} + \nu \\
\quad\quad\quad \mu^{+} \rightarrow e^{+} + \nu + \bar{\nu}
\end{array}
$$

Even though the A_2 is heavier than the proton, there is no baryon among its decay products. By our new definition it is a meson, something it would not be under the old scheme of things.

The concept of baryon and meson classifications is given a slightly more quantitative aspect by defining a quantity known as the *baryon number, B.* This is the number of protons that appear in the final state of a decay. For all of the baryons we have discussed above, $B = 1$, while for all of the mesons, $B = 0$. For antibaryons, $B = -1$.

Classification by Speed of Decay: Strange Versus Nonstrange

In the last few chapters we have seen how a series of hadrons can be produced in the laboratory and in cosmic ray experiments. All of these are created in a time scale characteristic of the strong interactions, but some of them seem to take an awfully long time to decay. Thus, the speed of decay provides another way of distinguishing among particles. The strange particles seem to decay in times on the order of 10^{-10} second. The nonstrange particles (which we have been calling resonances up to this point) decay in 10^{-23} second or so.

In 1953 two physicists, Murray Gell-Mann (then at the University of Chicago) and Kazuhiko Nishijima (at Osaka University in Japan) independently suggested a theory that seemed to provide a good way of thinking about this phenomenon. They reasoned that in the case of the two stable particles, the proton and electron, the infinite lifetimes could be thought of as being due to a conservation law. For the electron, there is no lighter negatively charged particle into which it can decay, so the law of conservation of charge tells us that the electron cannot decay. In the same way, the proton does not decay because of the conservation of baryon number, a law that we shall discuss later. Thus, long lifetimes seem to be associated with conserved quantities.

Could it be that "longish" lifetimes are associated with quantities that are almost (but not quite) conserved? Gell-Mann and Nishijima postulated that there was another quantity similar to electrical charge carried by every particle. Nishijima called this the η (eta) charge, while Gell-Mann called it *S.* For the nonstrange particles, *S* is zero. For the lambda, sigma, cascade, and *K*-meson families, however, *S* is not zero. For technical reasons, the *S*-charge of all of these particles is taken to be -1. In a process such as $\Lambda^\circ \rightarrow p + \pi^-$ we then have the *S*-charge changing from -1 on the left to 0 on the right. If the *S*-charge were conserved like an ordinary electrical charge, this decay would be absolutely forbidden and the lambda would be a stable particle. The fact that the lambda is not stable means that *S*-charge, whatever it is, is not conserved exactly. The fact that the lambda has a long lifetime, however, does mean that it is almost conserved.

The other systematics of the strange particles are also explained by this hypothesis. If we assign the cascade an *S*-charge of -2, for example, then the double slow decay chain

$$\begin{array}{l} \Xi^- \rightarrow \Lambda^\circ + \pi^- \\ \quad\quad\; \hookrightarrow p + \pi^- \end{array}$$

can be understood, since the *S*-charge changes by one unit in each step.

When a strange resonance decays quickly into a lambda and a pion, however, we say that both the resonance and the lambda have an *S*-charge of -1. Although the *S*-charge is nonzero before and after the decay, it does not change during the decay. This means that the reaction can proceed quickly, as indeed it does.

Since the nonstrange particles have $S = 0$ and the strange particles have $S \neq 0$, Gell-Mann called the new quantity *strangeness,* a name that has stuck. We say, therefore, that the Λ° has strangeness -1, the cascade strangeness -2, and so on.

Physicists readily adopted this whimsical turn of phrase because it tended to give a rather serious and abstract subject a lighthearted side. Perhaps in the aftermath of the Manhattan Project it was felt that a little humor would do the profession some good. This lightness has been bought at a price, however. Although physicists mean something precise when they talk of strangeness, and can relate it to measurable quantities, such as the length of tracks in a bubble chamber, it is almost impossible to disentangle the precise concept from the everyday connotations of the word. When a middle-aged savant speaks learnedly of baryons, hadrons, and unified field theories, he can sound impressive. When he talks of strangeness (or, more recently, of such concepts as "color" and "flavor"), he sounds slightly frivolous. The names tend to trivialize a rather serious endeavor, and, in the present climate of research funding, may prove a positive embarrassment to the field.

This is all water under the bridge at this point, though, since both the concept and the term strangeness are firmly entrenched as a property of elementary particles.

Classification by Internal Dynamics: Spin

ALTHOUGH it is an incorrect depiction, in this section we will think of elementary particles as small spheres of material rather than as smeared out wave functions. By so doing, we can see that there is one possible type of motion of the particle that we have not yet discussed—the possibility that the particle may rotate (or spin) around an axis. There are many examples of spinning spheres in the macroscopic world. Perhaps the most familiar is the earth itself, which turns on its axis once a day.

In describing such a rotational system, physicists like to define a quantity called the *angular momentum.* This is a quantity analagous to ordinary linear momentum (mass times velocity). Just as linear momentum is conserved and is related to the tendency of a moving object to keep moving unless acted on by a force, angular momentum is also conserved and expresses the tendency of a rotating object to keep on rotating unless a force acts to slow it down. The angular momentum of a spinning object is usually represented by an arrow; therefore, to define it completely we have to specify both the direction and length of the arrow.

In Illustration 42 we show two rotating spheres. They are identical in every respect except for the direction of rotation. We could express this difference by saying that one sphere is spinning clockwise and the other counterclockwise. It turns out, however, that it is more conve-

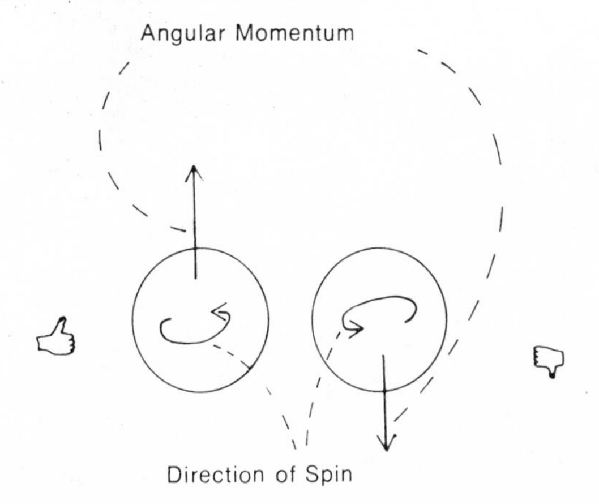

42. Angular momenta of two spheres.

nient to say that the angular momentum arrow representing the angular momentum for the two spheres points in a different direction for the two cases. By convention, the direction of this arrow is defined by something called the *right hand rule.* The rule says that if you wrap the fingers of your right hand in the direction of the rotation of the sphere, the angular momentum will point in the direction of the thumb of that hand. Thus, the angular momenta for the two spheres in the figure are as shown.

Since angular momentum is supposed to be related to the tendency of a body to keep rotating, we would expect the momentum to become larger as the mass and size of the body increase, and to grow smaller as the rate of spin is reduced. For a sphere of radius R and mass M, turning around every T seconds, the length of the arrow representing the angular momentum turns out to be

$$L = \frac{2}{5} MR^2 \cdot \frac{2\pi}{T}$$

For a classical macroscopic sphere, such as the earth, the period of rotation, T, can be any number at all. When we go to quantum mechanical objects, such as elementary particles, however, the situation is different. Just as in quantum mechanics an electron in an atom can only be at certain specified distances from the nucleus, the same laws state that

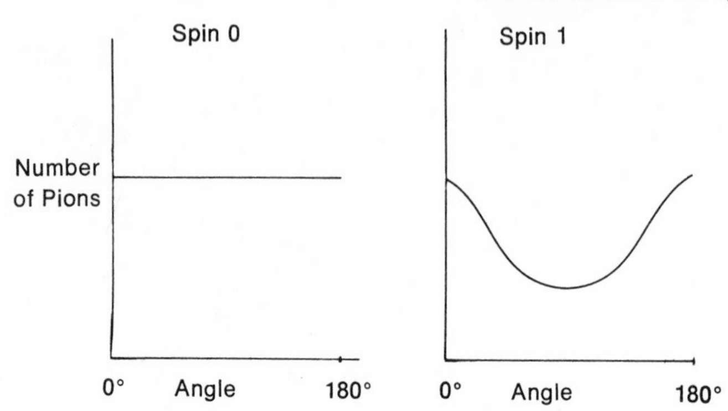

43. Graphs typical of spin zero and spin one particles.

a particle can spin only at certain specified rates. According to quantum mechanics, the angular momentum of a spinning particle can only have the values

$$L = J(J + 1)\frac{h}{2\pi}$$

where J is called the spin quantum number of the particle. J can have only half- or whole-integer values—that is, it can be 1/2, 1, 3/2, 2, 5/2, 3, and so forth. It cannot be 3/4 or 2/3 or any other value in between.

The spin of unstable elementary particles is usually deduced from looking at the directions in which the decay products are emitted when the particle disintegrates. The usual procedure is to construct a graph of the number of times a particular decay product (for example, a pi-meson) emerges at a particular angle with respect to the direction of motion of the original particle. Depending on the spin of the original particle, such graphs will have different (but well-defined) shapes. For example, graphs typical of spin zero and spin one particles undergoing decay are shown, above, in Illustration 43.

The spins of the particles we have encountered so far are summarized in the following table:

Spin 0

pion

Spin $\frac{1}{2}$

electron, proton, neutron, muon, neutrino, Λ°, Σ, Ξ,

Spin 1

photon, ρ

Spin $\frac{3}{2}$

Δ

The highest spin particle found so far is called the *h*-meson. It has a mass of 2,040 MeV and a spin of four. There is, however, no theoretical limit to the spin—it can be as high as it wants. Consequently, the fact that very high spin particles have not been found is probably related more to the lack of interest in designing experiments to find them than to any intrinsic natural limitation.

There is still another important difference between the classical and quantum mechanical properties of particles with angular momentum. If we define a direction in space (for example, by putting the particles in a magnetic field that points in a fixed direction), there is no necessary connection between the spin of a classical object and this direction. The situation is like the one pictured on the left of Illustration 44; that is, the classically spinning particles can have their angular moment pointing

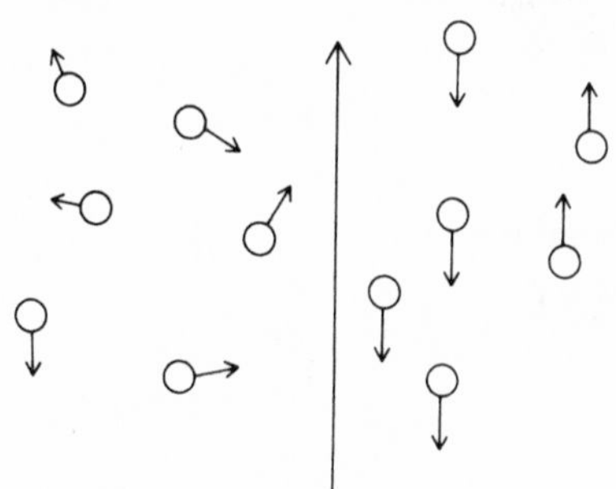

44. Angular momentum of classical vs. quantum mechanical particles.

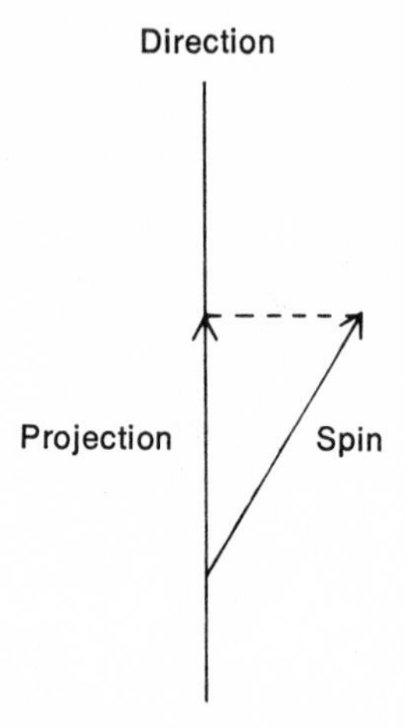

45. The geometry involved in projecting a spin onto an axis.

in any direction at all. The quantum mechanical case, however, is like the situation pictured on the right (in Illus. 44): The angular momentum of quantum mechanical particles can only point in certain specified directions in space.

The allowed directions for the spin of a quantum mechanical particle are determined in the following way: A direction in space is chosen, and then the projection of the spin along this direction is determined. This process is shown, above, in Illustration 45. The allowed directions of the spin are determined by the requirement that the projection of the spin have one of the following values:

$$J \frac{h}{2\pi}$$

$$(J - 1) \frac{h}{2\pi}$$

$$(J - 2) \frac{h}{2\pi}$$

$$\cdots$$

$$(-J) \frac{h}{2\pi}$$

The number of directions in which a spin can point thus depends on the magnitude of the spin itself. For spin 1/2 particles, there are only

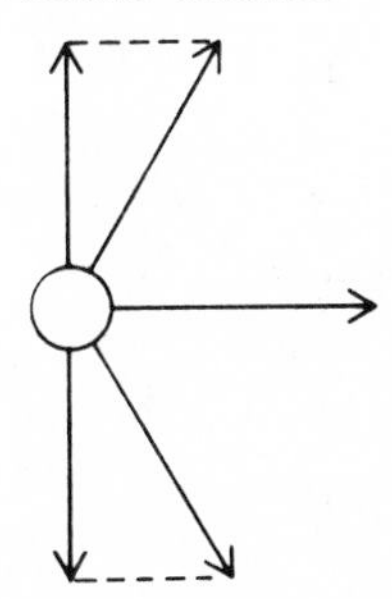

46. The allowed orientations for the angular momentum of a spin one particle. The spins are shown on the right in their three orientations, and the projections of the spins on the left are (from top to bottom) $+1$, 0, and -1.

two—"up" and "down." For spin one particles, there are three—"up," "down," and "sideways." The number becomes higher as the spin increases. Illustration 46 shows the allowed spin orientations.

Finally, we note that this requirement tells us that the number of allowed orientations of a particle of spin J is given by $N = 2J + 1$. For example, for spin one the allowed orientations are those corresponding to projections of 1, 0, and -1, respectively.

Classification by Electrical Charge: Isospin

WE have seen that one of the most important properties of a particle is its electrical charge. We have also seen that many of the elementary particles seem to occur in families where the members are identical in all respects—even to the point of having roughly equal masses—except in electrical charge. The pion, coming in positive, neutral, and negative varieties, is a good example of this kind of occurrence. Furthermore, it appears that as far as the strong interaction properties of a family of particles are concerned, it makes very little difference which member of the family is involved. The strong interactions, in other words, do not seem to depend on the electrical properties of an individual particle.

In an attempt to provide a simple unified way of understanding these facts, theoretical physicists have drawn an analogy between the laws that govern spin (see the previous section) and the electrical properties of particle families. Suppose, they say, that the laws that govern ordinary spin also govern another kind of quantity—a quantity related to the charge. To stress this purely mathematical analogy, this new quantity is called *isotopic spin.* Suppose further, they say, that there is some abstract, mathematical space in which we can imagine the isotopic spin to be an arrow, just as we represented ordinary spin as an arrow in the last section. In this abstract isotopic-spin space the isotopic spin can therefore only be oriented in certain directions, just as the ordinary spin can only be oriented in certain directions in ordinary space. This means, for example, that an isotopic spin of one would correspond to a situation in the abstract space like the one we have pictured in Illustration 47. The projections of this spin along the axis are $+1$, 0, and -1.

Do we know of any family of particles in which the charge comes in three states, positive, zero, and negative? The answer, of course, is yes. The pi- and rho-mesons both qualify. We then agree that instead of talking about these particles as coming in different charge states, we shall regard them as a single particle that can have three possible orientations of its isotopic spin vector. In this way of looking at things, the members of any family are now considered (at least as far as the strong interactions are concerned) the same particle. The differences between them are thought of purely as differences in the orientation of the isotopic spin, and we would no more say that this made them different types of particles than we would claim that two electrons were different

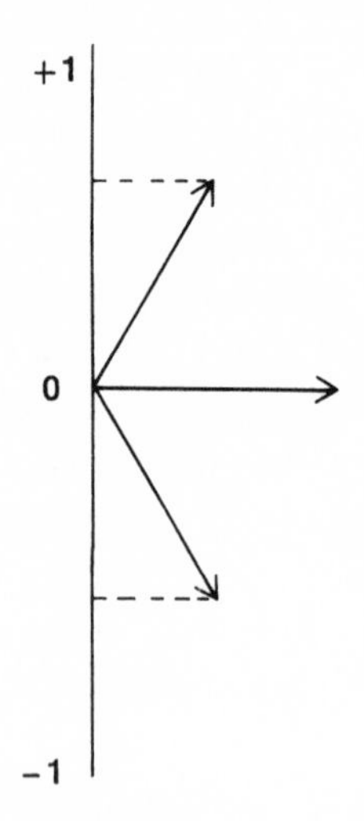

47. Possible orientations of isospin vector for $I \doteq 1$ has positive.

kinds of particles just because their angular momenta pointed in different directions.

Because the number of different orientations of an isotopic spin vector is given by $N = 2I + 1$, where I is the isotopic spin, the number of members of a family can be used to determine the isotopic spin directly.

The connection between electrical charge and isospin projections, in general, is given by the equation $Q = I_z + B/2 + S/2$, where I_z is the projection of the isotopic spin, B is the baryon number, and S is the strangeness. For the rho- and pi-mesons, both B and S are zero; hence, nothing in the above argument is changed.

We can look at the proton and neutron as examples of how isotopic spin works. Here are two particles in the same family; therefore, we have $2 = 2I + 1$ or $I = 1/2$. An isotopic spin of $1/2$ will have two projections, $+1/2$ and $-1/2$. For the nucleons, $B = 1$ and $S = 0$, so the two possible charges are $Q = 1/2 + 1/2 = 1$ and $Q = -1/2 + 1/2 = 0$. Thus, the proton corresponds to the $+1/2$ orientation of the isotopic spin and the neutron to the $-1/2$ orientation.

For the delta, on the other hand, there are four charge states, so $I = 3/2$. Since $B = 1$ and $S = 0$, the possible charge states are

$$Q = \frac{3}{2} + \frac{1}{2} = 2$$

$$Q = \frac{1}{2} + \frac{1}{2} = 1$$

$$Q = -\frac{1}{2} + \frac{1}{2} = 0$$

$$Q = -\frac{3}{2} + \frac{1}{2} = -1$$

In the following table the results of this kind of calculation are presented for some of the particles we have studied:

PARTICLE	N	I	B	S	Q
π	3	1	0	0	$+1, 0, -1$
ρ	3	1	0	0	$+1, 0, -1$
p, n	2	$\frac{1}{2}$	1	0	$+1, 0$
Δ	4	$\frac{3}{2}$	1	0	$+2, +1, -1, 0$
Λ	1	0	1	-1	0
Σ	3	1	1	-1	$+1, 0, -1$
Ξ	2	$\frac{1}{2}$	1	-2	$0, -1$

One point that should be emphasized before we move on is that the families that are described by isotopic spin exist only among the hadrons; therefore, we do not speak of the isotopic spin of leptons. Furthermore, although we have talked about only one use of isotopic spin (its ability to simplify our picture of particle families), there are many predictions about reactions that can be made under the assumption that the laws governing ordinary spin also govern isotopic spin. Since these predictions are invariably correct, there is a good deal of experimental evidence to support the rather abstract presentation given here.

Other Methods of Classification

THERE are a few other properties of elementary particles which, while not of great importance for the process of classifying, become important in the discussion of conservation laws. We will present them briefly here.

Parity. In Chapter III we saw that every quantum mechanical particle can be described by a wave function that is ultimately related to the probability that a measurement would find the particle at a particular point in space. The parity of the particle has to do with what happens to the wave function when a particular kind of mathematical operation is performed. The operation is essentially the interchanging of right and left—something like looking at the wave function in a mirror. In technical terms, the value of the old wave function at a point x is replaced by the value of the wave function at the point $-x$. In the wave function

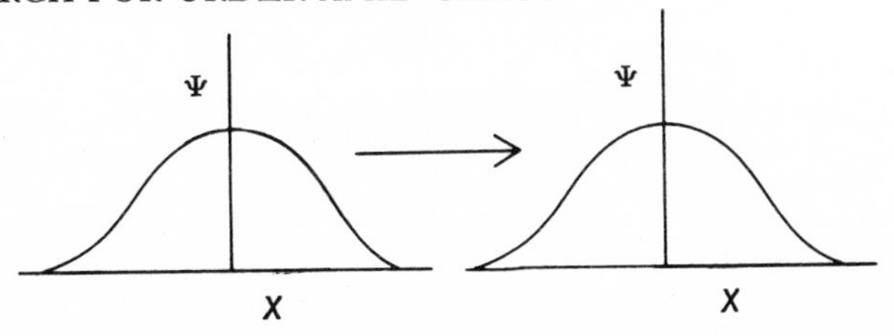

48. A wave function that is even under parity.

in Illustration 48, performing this operation (or, equivalently, looking at the wave function in a mirror) just reproduces the original wave function. The particle described by this wave function is said to have positive parity.

On the other hand, in Illustration 49 we show a wave function for which this process produces a new wave function that is precisely the negative of the old one. Such a function is said to be odd under parity, and the particle it describes is said to have negative parity.

Charge Conjugation. If we imagined taking a collection of particles and changing every one of them into its antiparticle, we would have performed the operation known as charge conjugation. In general, there is no particular relation between the wave function of a particle and that of its antiparticle, but for a few neutral particles, such as the π°, the operation will produce a wave function that is either plus or minus the original one. Thus, as with parity, we can speak of positive or negative charge conjugation for this limited class of particles.

Time Reversal. Time reversal is not, strictly speaking, a property of particles, but we include it here for completeness. If you watch an

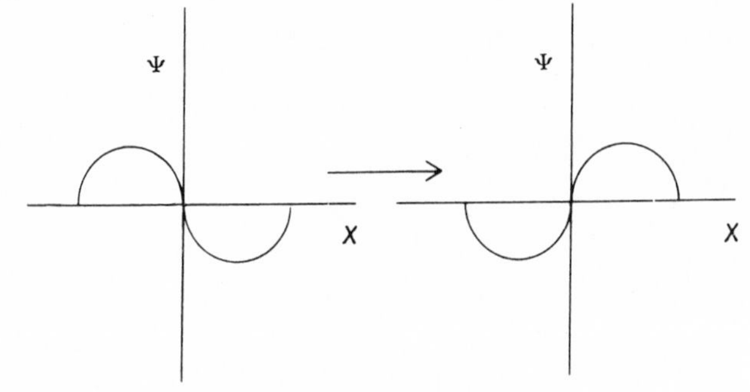

49. A wave function that has negative parity.

interaction involving elementary particles, you can imagine putting the whole process on film. The operation of time reversal corresponds to running the film backwards. This operation does not change a particle into plus or minus itself, but it obviously has an effect on what is seen. To take one example, a particle that is spinning so that its angular momentum is "up" will, upon time reversal, be seen to have spin "down."

A Summary

TO classify a new elementary particle we therefore have to ask first whether it is a hadron or a lepton. If it is the latter, there is not much else to do. If the former, we proceed to see if it is a baryon or meson, strange or nonstrange, and to ascertain its spin, isotopic spin, and, if applicable, its charge conjugation. This information, together with the mass of the particle, gives us all of the properties we are likely to need to know. In the following table we present this information for all the particles discussed so far.

SUMMARY OF PARTICLE PROPERTIES

	PARTICLE	MASS	J	S	I	P	B
Photon	γ	0	1	0			0
Leptons	e	0.51	$\frac{1}{2}$	0			0
	μ	105	$\frac{1}{2}$	0			0
	ν	0	$\frac{1}{2}$	0			0
Mesons	π	140	0	0	1	−	0
	K^-	494	0	−1	$\frac{1}{2}$	−	0
	K°	498	0	−1	$\frac{1}{2}$	−	0
	ρ	770	1	0	1	−	0
Baryons	p,n	938	$\frac{1}{2}$	0	$\frac{1}{2}$	+	1
	Λ	1,115	$\frac{1}{2}$	−1	0	+	1
	Σ	1,190	$\frac{1}{2}$	−1	1	+	1
	$\Xi^{\circ,-}$	1,318	$\frac{1}{2}$	−2	$\frac{1}{2}$	+	1
	Δ	1,232	$\frac{3}{2}$	0	$\frac{3}{2}$	+	1
	$\Sigma(1385)$	1,385	$\frac{3}{2}$	−1	1	+	1

CHAPTER IX

The Road to the Quark Model

The quark model is to physics
as the folk song is to music.

—ANONYMOUS

Some Systematic Relationships Among Particles

WHEN the first resonances were discovered, the general feeling among physicists was that they were somewhat anomalous and special. This feeling was reflected in the names given them. What we have called the delta, for example, was called the N*, the assumption being that this would be *the* pion nucleon resonance. Similarly, the strangeness −1 baryon resonance was called the Y*. But as the number of known resonances began to increase dramatically in the 1960s, this nomenclature became more and more unwieldy. Other resonances in the pion-nucleon system began to appear and were christened the N** and the N***. As Matt Roos said in his first review of elementary particles, "We expect the starred notation to become unpopular by the time a resonance is discovered which needs eight stars."

A look at the pion-nucleon resonances will serve to introduce the modern nomenclature for resonances and to illustrate some of the regularities that are seen in them. The 1976 *Review of Particle Proper-*

ties lists an even dozen resonances, which have the following properties: $B = 1$, $S = 0$, and $I = \frac{1}{2}$. All these particles decay into a nucleon and a set of pi-mesons. Consequently, we would be justified in calling all of them by the name N*. In addition, the quantum numbers listed above are exactly those we have described for the nucleon itself. The other properties of these particles (such as mass, parity, and spin) vary as we go through the list. The convention that has been adopted is to denote them all by the letter N and to add in parentheses the mass measured in MeV. For the nucleonlike resonances, the list includes:

N(1470)	N(1780)
N(1520)	N(1810)
N(1535)	N(2190)
N(1670)	N(2220)
N(1688)	N(2650)
N(1700)	N(3030)

Since 1976, more particles have been added to this list.

The decay of all these particles takes place primarily through the creation of a nucleon and a single pion. However, all of them can decay by other processes, such as

$$\mathrm{N}(\quad) \rightarrow \Delta\pi$$

or

$$\mathrm{N}(\quad) \rightarrow Np$$

In fact, the decay is very similar to the process by which light is emitted from an atom (see Chapter I). When an electron is in a high orbit in an atom, it can move into lower orbits in a number of ways. It can make one big jump (emitting a high-energy photon), or it can make a series of smaller jumps through intermediate orbits (emitting a series of lower energy photons).

In the same way, it appears that these nucleon resonances can decay to a nucleon plus pions through a single transition, or they can decay in a series of steps. This similarity with the electron is the first hint we have that under the avalanche of elementary particles we might still be able to find an underlying simplicity. Maybe calling all the resonances in the above list separate particles is as pointless as insisting that two atoms are fundamentally different because their electrons are in different orbits.

The nucleon is not the only low-mass baryon that has a "family." We could compile a list just like the one above for nonstrange baryons with isotopic spin 3/2. This would be the delta family, and the lowest mass member of the family, the Δ(1236), is the familiar resonance we have been calling the Δ. Other families are listed in the table below. One member of the sigma family, the Σ(1385), was the first strange resonance, whose properties we discussed in Chapter VII.

B	S	I	NAME OF FAMILY	NUMBER KNOWN IN 1976
1	-1	0	Λ	10
1	-1	1	Σ	10
1	-2	$\frac{1}{2}$	Ξ	3

There are fewer known meson resonances than resonances associated with the baryons. This fact is related more to the problems involved in doing a complicated phase space analysis than to any fundamental differences between mesons and baryons. We shall return to this point later, but what evidence we have indicates that there are meson families similar to the baryon families we have just enumerated.

Thus, by using the classification schemes outlined in Chapter VIII we can apply an enormous conceptual simplification to the problem of hadron proliferation. If each new resonance is regarded as a new member of a family, having no more fundamental significance than a new electron orbit in an atom, we have already succeeded in bringing a large measure of order into a previously chaotic world. In this way of looking at things, the number of families is rather small, even though the number of particles is large.

The Eightfold Way

THE first really successful scheme for showing the fundamental connection between particles in different families was developed independently in 1961 by Murray Gell-Mann at the California Institute of Technology and Yuval Ne'eman, who at the time was filling the roles of physicist at Imperial College, London, and military intelligence attaché at the Israeli embassy. This scheme bears the same logical relation to elementary particles as the periodic table does to chemical elements. Perhaps if we expand on this analogy a bit it will help us to understand what was done.

When the number of known chemical elements began to approach the hundred mark during the last century, it was recognized that some sort of order would have to be discovered among the proliferating elements. The Russian chemist Dmitri I. Mendeleev noticed that if the chemical elements were arranged in rows so that the atomic weight increased from left to right, and if the number of elements in the rows was adjusted properly, then elements in the same column would have similar chemical properties. Thus, a connection between atomic weight and chemical properties was established. In addition, the few "holes" in the table—places where elements should have been but weren't—led to the discoveries of scandium and germanium, two elements that were unknown up to that time.

In considering the periodic table, however, it is extremely important to realize that if someone had asked Mendeleev why the first row contained two elements while the next contained eight, he would not have been able to answer. The periodic table worked, but until the advent of quantum mechanics in the twentieth century no one understood why.

This same spirit of order-without-explanation led to the development of the periodic table of the elementary particles, which has been called the *eightfold way* (because it predicts that many hadrons will be grouped in sets of eight) and SU(3) (a technical mathematical term describing the properties of the groupings). Perhaps the best way to understand this table is to plunge right in and start ordering the particles.

Suppose we make a graph in which the vertical axis is the quantity $B + S$ for any particle, and the horizontal axis is I_z the projection of the isotopic spin. On such a graph (see Illus. 50), any particle will be represented by a point. For example, in the plot in the illustration we show the Σ baryons. All of them have $B = 1$, $S = -1$, so that $B + S$ (a

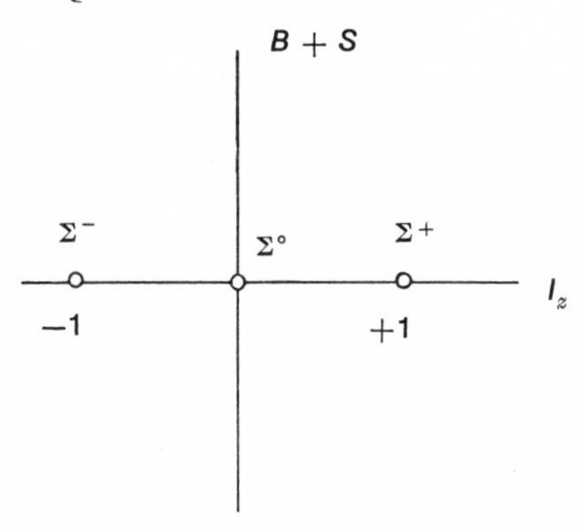

50. The sigma baryons on a graph such as that discussed.

quantity sometimes called the hypercharge) is zero. At the same time, recalling that

$$I_z = Q - \frac{B + S}{2}$$

the three charge states of the Σ give $I_z = 1, 0$, and -1, respectively. Consequently, on such a graph, the Σ family appears as three points spaced out along the horizontal axis.

Referring to the table at the end of Chapter VIII, we see that the nucleon, the sigma, the lambda, and the cascade all have the same spin, baryon number, and parity, and have masses that are roughly equal. What if we plotted all of these particles on a graph such as that shown above? You can test your facility with the concepts we have introduced so far by verifying that you will get a set of points like the ones shown in Illustration 51. In this graph the sigma and the lambda are placed apart from each other so that they can be easily seen, but both are at the point $(B + S) = 0, I_z = 0$.

The eight particles in the graph (Illus. 51) are similar in many respects. It turns out that grouping them in this way tells us a great deal about the way they contribute to the strong interaction. For example, if we read across the rows in the graph, all the particles have the same strangeness. If we read along a diagonal from upper left to lower right, they all have the same electrical charge. The grouping shown is thus

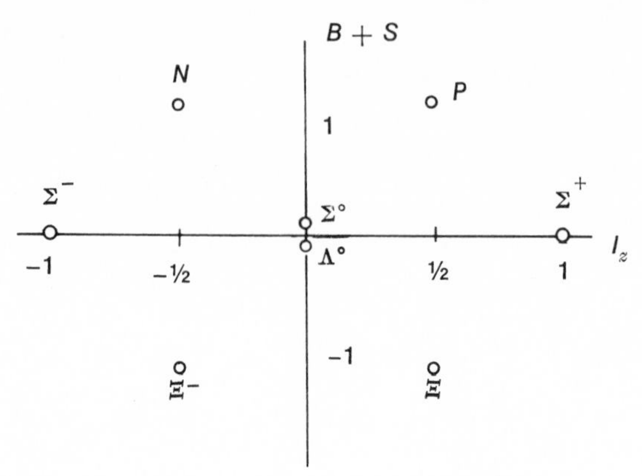

51. The baryon octet.

similar to the periodic table of the chemical elements—it tells us about the way the particles behave, but does not tell us why.

All the hadrons we have discussed so far can be clustered in groups of particles with equal spin, baryon number, parity, and (roughly) equal masses. The mathematical formalism tells us how many should be in each group. With a few exceptions, the lower mass particles we have been discussing fall into groups of eight, as the baryons pictured in Illustration 51 do. Such a grouping is called an octet, and the origin of the term eightfold way is now obvious. For reference, we show a meson octet in Illustration 52, on page 137. The η (eta) meson (which we have not mentioned previously) has spin zero and mass 549 MeV. The K^+ and $\overline{K}^\circ$ are the antiparticles of the K^- and K°, respectively.

During the early 1960s, it became clear that grouping the elementary particles as suggested by the eightfold way was an enormously fruitful way to think about them. The underlying mathematical formalism (a branch of modern mathematics called group theory) allowed theoretical physicists to relate the lifetimes of different resonances to each other and to predict the differences in the masses of particles within a given grouping. Being able to *predict* the lifetime of a resonance that previously had to be accepted from experiment meant that the amount of arbitrariness in the properties of the particles was being reduced. Instead of having to accept the new particles as they showed up in experiments, physicists could now begin to predict where these particles ought to be seen and to predict some of their properties. (We will see

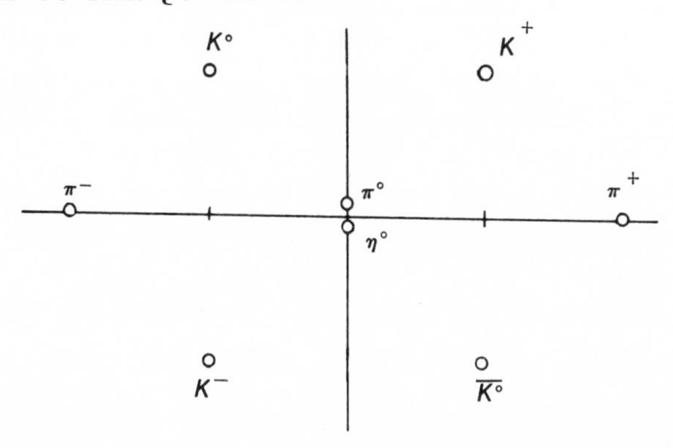

52. The meson octet.

later how one of these predictions was obtained.) Like the periodic table, the eightfold way led to an ordering of the elementary particles, an ordering that significantly reduced the previous chaos.

Yet there is still cause for an underlying uneasiness about the whole procedure. We saw that Mendeleev could not answer the question about why the periodic table should be arranged as it was. In the same way, we cannot say why particles should be grouped on a graph like the ones shown above. Simply stated, why should the graph have $B + S$ on one axis and I_z on the other? Why not some other combination of quantum numbers? In mathematical terms, why is it SU(3) and not some other symmetry that operates in the elementary particle world? Just as a full understanding of the chemical elements had to wait for the development of quantum mechanics, a deeper understanding of the eightfold way had to wait for a few years until a new simplification took place in our way of looking at the hadrons.

The Quark Model

IN 1964, Murray Gell-Mann at the California Institute of Technology and George Zweig in Geneva, Switzerland, independently suggested a simple physical explanation for the success of the eightfold way. Just as the periodic table could be explained once it was accepted that the atom possessed a structure and that electrons moved around a nucleus in orbits, the tendency of elementary particles to form octets could be understood if they, too, were made up of some still more elementary constituents. Hence, the question that had to be asked was, "What sort of constituents would be necessary to produce the elementary particles as we know them?"

The answer turns out to be rather surprising. In the original picture —also based on group theory and SU(3)—there were three constituents making up all the hadrons. These are called *quarks* (from a line in James Joyce's *Finnegans Wake*—"Three quarks for Muster Mark"). In order to produce the known elementary particles, the quarks must have unusual properties. For example, they have to have electrical charges that are fractions of the charge on the electron and proton. This alone would make them unique, since every other known electrical charge is either equal to one of these or some whole number of them.

On the other hand, no matter how unusual these new particles might be, there is plenty of circumstantial evidence indicating that the particles we have been calling elementary up to this point are, like the atoms before them, made up of constituents. The existence of particle families certainly suggests this, as does the unexplained ordering of the particles brought about by the eightfold way. Thus, no matter how unusual the quarks might be—either in name or in properties—it is worth spending time to see what the evidence for them is.

In the table we present the relevant properties of the three original quarks (the names are those now in use, rather than those originally proposed). For each of these there is a corresponding antiparticle—the antiquark—with opposite charge, baryon number, and strangeness. The antiquarks are denoted by the symbols $\bar{u}$, $\bar{d}$, and $\bar{s}$.

NAME OF QUARK	SYMBOL	J	Q	S	B	I	I_z
Up	u	$\frac{1}{2}$	$\frac{2}{3}$	0	$\frac{1}{3}$	$\frac{1}{2}$	$\frac{1}{2}$
Down	d	$\frac{1}{2}$	$-\frac{1}{3}$	0	$\frac{1}{3}$	$\frac{1}{2}$	$-\frac{1}{2}$
Strange	s	$\frac{1}{2}$	$-\frac{1}{3}$	-1	$\frac{1}{3}$	$\frac{1}{2}$	0

The quark names are not so unusual as they may seem. The u and d quarks form a family of isotopic spin 1/2, and "up" and "down" refer to the two projections of that spin. "Strange" refers to the fact that this quark has $S = -1$, while the others have $S = 0$.

How could we go about constructing the hadrons from these building blocks? Let us start with the baryons. These particles all have $B = 1$.

Since all of the quarks have $B = 1/3$, we can deduce an important principle of the quark model: *Baryons are made from three quarks.* The values of B for the three constituent quarks in the baryon will therefore add up to $B = 1$, as they should.

To find which quarks go with which baryons, we have to see which combination will give the correct spin, charge, and strangeness. Consider the delta baryon of charge $+2$ as an example. This is a particle with charge $+2$, spin $3/2$, and strangeness 0. We see immediately that there can be no s quarks in the Δ^{++}. A single s quark would give the particle $S = -1$, while a combination of s and $\bar{s}$ (which would give $S = 1 - 1 = 0$) would give $B = 1/3$ (remember that the antiquark has $B = -1/3$).

Is there any way we can combine the u and d quarks to make a Δ^{++}? There is only one way to do this and still have a charge $+2$, and that is to have three u quarks. The total charge of this combination will be $Q = 2/3 + 2/3 + 2/3 = 2$. Similarly, to get a total spin of $3/2$ from three objects of spin $1/2$, we have to have all three spins lined up, pointing in the same direction. Thus, in terms of quarks, the Δ^{++} must look something like Illustration 53, below. In the same way, the Δ^{+} must also have three spins lined up, but must have two u quarks and a d quark to give total charge $Q = 2/3 + 2/3 - 1/3 = 1$. It will look something like Illustration 54. The proton, on the other hand, must have spin $1/2$ and charge 1. Consequently, it must be made up of two u quarks and a d quark, but the spins must be arranged differently from the Δ^{+}. A possible arrangement for the proton is shown in Illustration 55.

We can now see the essential point of the quark model. Up to this point we have had to regard the delta and the nucleon as two separate particles. All of the regularities we have discussed have not altered that essential fact. In terms of quarks, however, we see that there is a fundamental connection between these two particles. They are both made of the same kinds of quarks, but these quarks are arranged differently. The philosophical implications of this fact will be discussed later, but for the moment we note that it gives us a very simple picture of the process by which one particle changes into another.

Consider the fast decay $\Delta^{+} \rightarrow p^{+}\pi^{\circ}$. In terms of particles, we have to think of this decay as something that can be seen in the laboratory but not understood. In terms of quarks, however, we can picture this event as shown in Illustration 56. The d quark flips its spin from up to down,

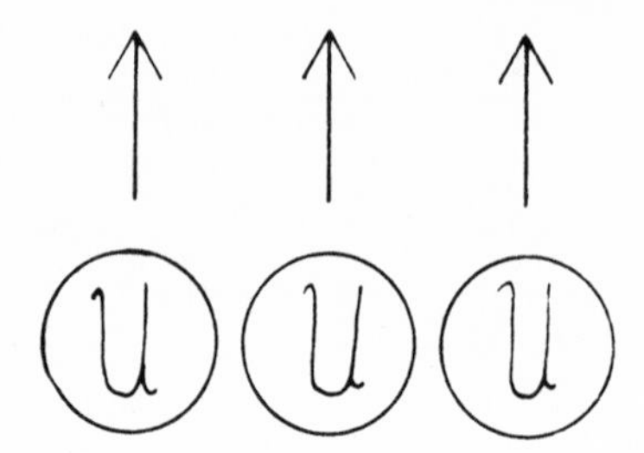

53. Δ^{++} in terms of quarks.

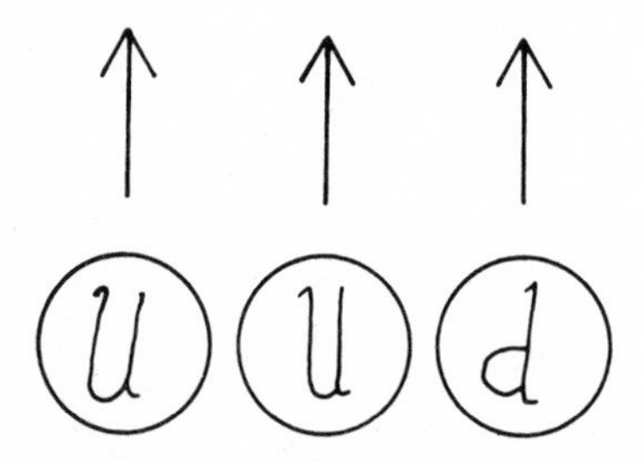

54. Δ^{+} in terms of quarks.

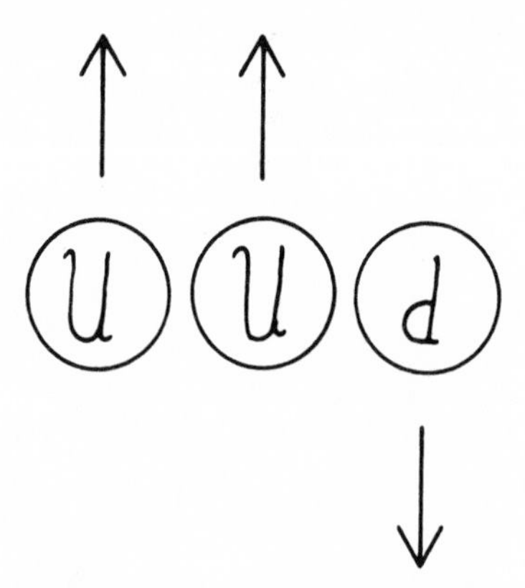

55. Possible arrangement for the proton, in terms of quarks.

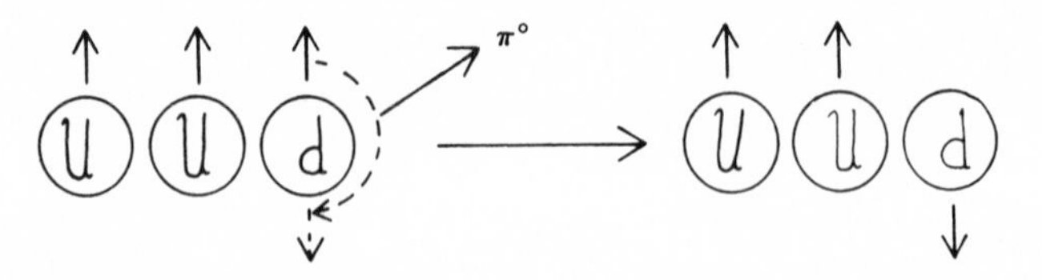

56. Fast decay, in terms of quarks.

emitting a neutral pion in the process. Since this sort of phenomenon happens all the time in nuclei (the pion is simply replaced by an X ray), physicists felt very much at home with it. Not only that, but once a simple picture like this is produced, it is possible to use the techniques of nuclear physics to calculate lifetimes of elementary particles, including lifetimes of particles in different octets. In the mid-1960s, many such calculations were made and, considering the extreme simplicity of the model, they were quite successful.

The quark model also gives us a way of understanding the fast-versus-slow decay systematics that first became apparent in cosmic ray experiments. To see how this works, let us consider the positively charged sigma (Σ^+). This particle has $S = -1$, so it must contain one s quark. In order to make up the charge of $+1$, the other two quarks must be u quarks. To get spin $1/2$, we must have something like what is shown in Illustration 57, below.

Now consider the decay $\Sigma^+ \rightarrow p + \pi^\circ$. In order for this decay to proceed, the s quark in the Σ^+ must change not its spin, but its identity. It must transform into a d quark in a process similar to that shown in Illustration 58.

It is not too unreasonable to suppose that it takes longer for a quark to transform itself into another quark than to flip its spin. The quark model rule for decays is thus: *A decay can proceed quickly only if the quarks involved do not have to change their identity.*

We can use this rule to discuss the decay of the $\Sigma(1385)$, a strange resonance that decayed quickly into a lambda and a pion. This particle has $S = -1$, $B = 1$, and spin $3/2$. If we consider the positively charged $\Sigma(1385)$, it must have a quark content like the Σ^+, except that the three quark spins must be aligned. It must, in other words, look similar to the representation in Illustration 59. If we consider the decay $\Sigma^+(1385) \rightarrow \Sigma^+ + \pi^\circ$, then it must proceed as pictured in Illustration 60. The similarity between this spin-flip process and the spin flip that allowed the delta to decay to a proton cannot be overemphasized. The processes are identical except for the type of quark that is flipping its spin. Both are fast because they involve only a spin flip.

With this rule, we can also understand the double slow decay of the cascade. An $S = -2$ particle must have two s quarks in it. The cascade must therefore look similar to Illustration 61. In order for this particle to decay, the two strange quarks must change to ordinary d quarks. This

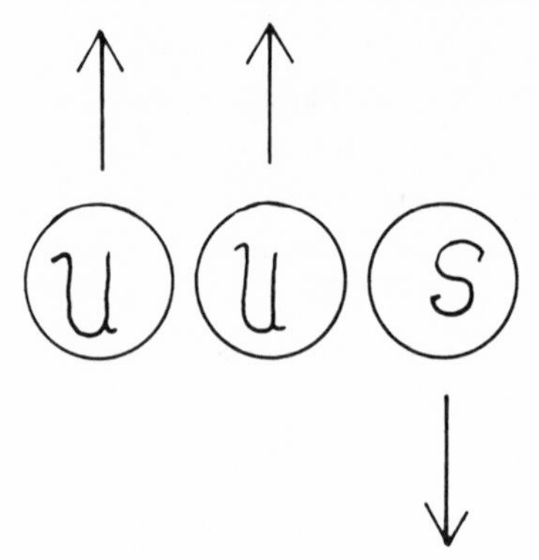

57. The Σ^+, in terms of quarks.

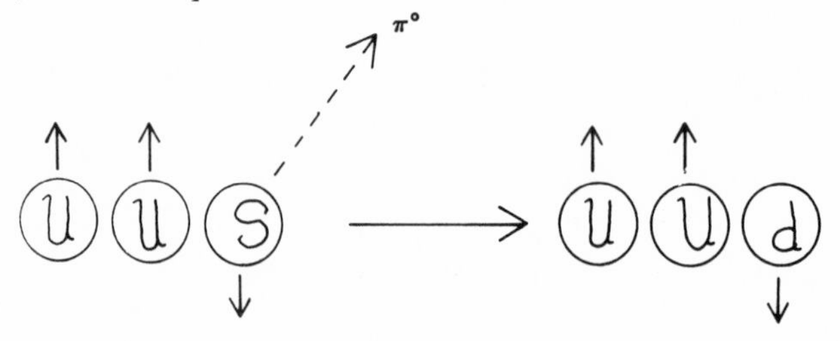

58. The *s* quark is a Σ^+ transforming into a *d* quark.

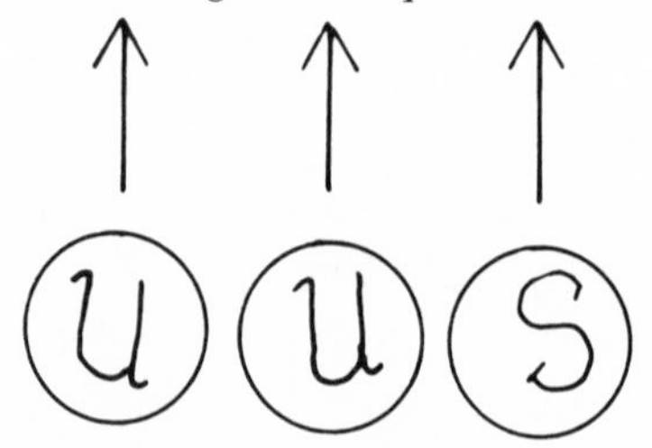

59. The $\Sigma^+(1385)$ particle, in terms of quarks.

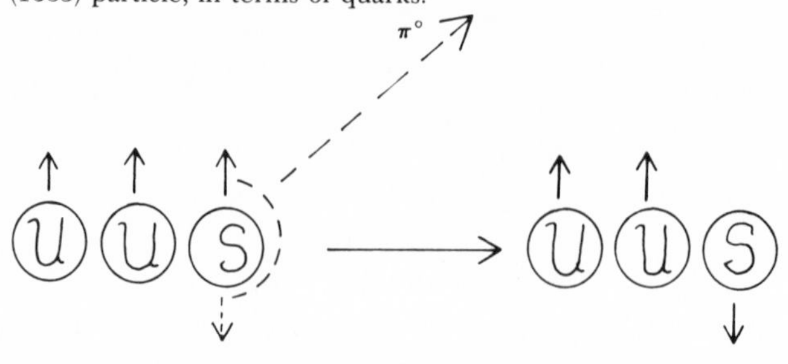

60. The decay process of $\Sigma^+(1385)$, in terms of quarks.

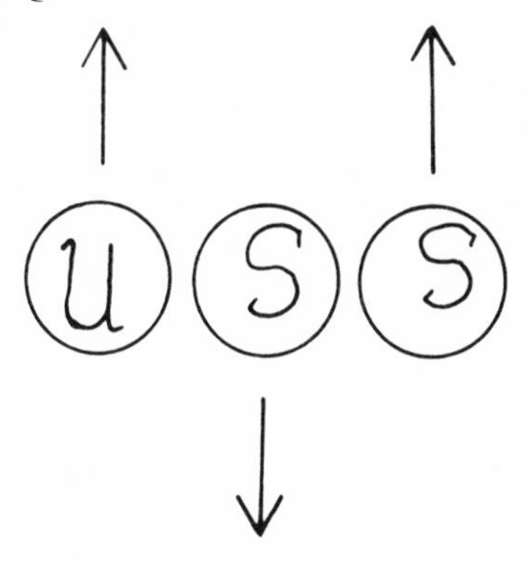

61. The cascade, in terms of quarks.

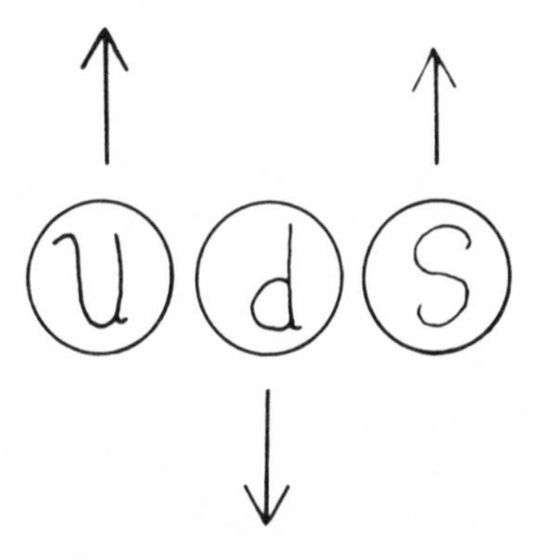

62. The Λ°, in terms of quarks.

happens in two steps. First one of them changes while emitting a neutral pion, which leaves a particle similar to that represented in Illustration 62. It has 0 charge, spin 1/2, and strangeness -1. This, of course, is the Λ°. This new particle then decays slowly by converting the remaining *s* quark into a *d* quark. Thus, we see that double slow decays correspond to situations in which we have to change two *s* quarks into two *d* quarks to get to the final state, rather than changing just one *s* to one *d*.

One caveat to this decay rule should be made at this point. In the decay $\Delta^{+} \rightarrow n + \pi^{+}$ we have a process similar to that shown in Illustration 63. A *u* quark flips its spin, emits a pion, and becomes a *d* quark. Doesn't this process involve a change in identity?

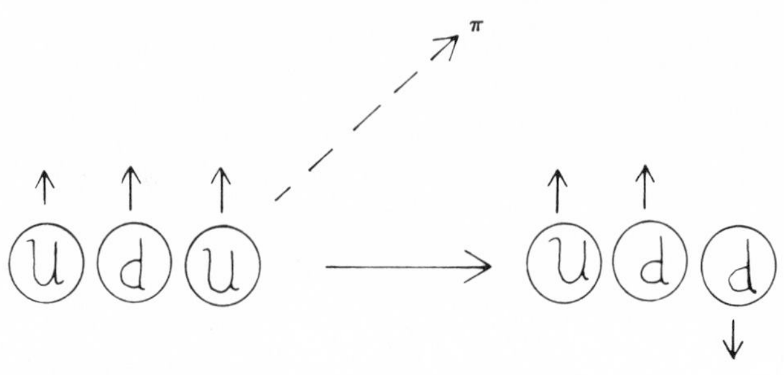

63. The decay of the Δ^+.

If you recall that the *u* and *d* quarks are members of the same isotopic spin family, you will see that the change in identity in this process is just another spin flip. The isotopic spin changes from up to down in isotopic spin space just as the ordinary spin changes from up to down in ordinary space. Consequently, changes between *u* and *d* quarks are on the same logical footing as spin flip, and do not constitute change of identity as we are using the term here.

The quark model also provides a way of thinking of the high-mass particle families discussed earlier in this chapter. All of the particles we have discussed so far involve three quarks that are stationary with respect to each other. This is certainly the simplest possible configuration for them, but the question naturally arises as to whether we could have a situation like that shown in Illustration 64, where the quarks are in orbits around each other. The answer to this question is yes. Not only are such states possible, but they account for many of the dozen nucleon resonances we listed earlier. Each resonance corresponds to a different

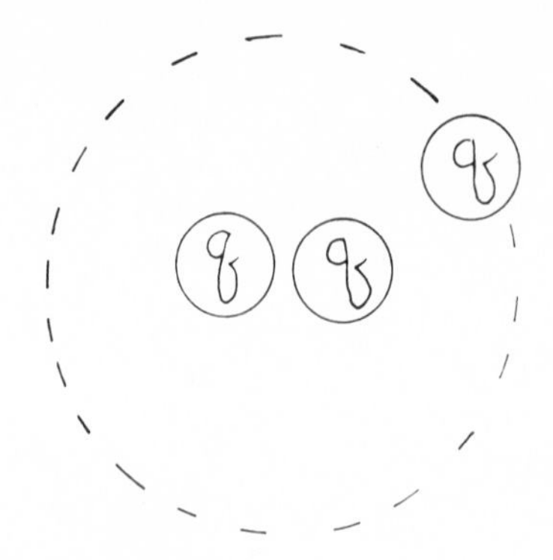

64. Quarks orbiting around each other.

set of orbits for the quarks. Since these orbits can get bigger and bigger, there is no limit, in principle, to how many resonances there can be. In this sense, it is a good thing the quark model was formulated when it was. Otherwise, we would now be absolutely flooded with particles.

The Quark Model for Mesons

IN the previous section we treated mesons as particles that are emitted when quark spin flip occurs. Actually, we know that mesons are hadrons and therefore must also be composed of quarks. The quark structure for mesons will be discussed in this section.

By definition, mesons have baryon number 0. The only combination of quarks that will produce this result is a quark plus an antiquark. For them, we will have $B = 1/3 - 1/3 = 0$. This leads to another rule for the quark model: *Mesons are made from a quark and an antiquark.*

Using this rule, we can find the right combinations for the different mesons just as we did for the baryons. Take the positive rho-meson as an example. This is a meson with $S = 0$, $Q = 1$, and $J = 1$. If there are to be any s quarks in the particle, they would have to appear in an $s\bar{s}$ combination. Such a combination could not have an electrical charge. Consequently, the ρ^+ must be made of u and d quarks along with their antiquarks.

Referring to the table of quark properties, we see that a u quark (charge 2/3) and a $\bar{d}$ antiquark (charge 1/3) will give us the proper charge for the ρ^+. It must therefore look something like Illustration 65. A similar argument shows that the π^+ must be as shown in Illustration 66. As was the case with the proton and the positive delta, the only difference between the two particles is in the alignment of quark spins.

In our discussion of the decay of the baryons, we treated the pions that were emitted as single particles. Clearly, we cannot treat the process $\rho^+ \rightarrow \pi^+ + \pi^\circ$ in the same way. We have to look a little more closely

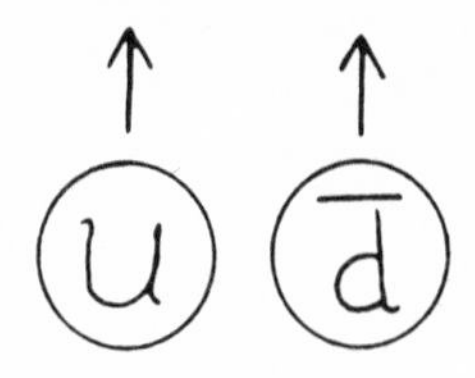

65. The ρ^+, in terms of quarks.

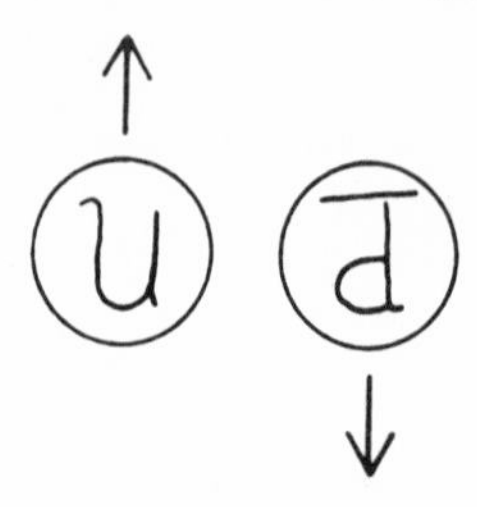

66. The π^+, in terms of quarks.

at the process by which a quark emits a pi-meson.

For this discussion, we can consider the quark that is going to be active during the decay process as being isolated in space, and that the other constituents of the particle will play the role of observer in whatever happens. In the simplest case, it should make no difference whether the quark we are looking at is attached to an antiquark (and, hence, is in a meson) or to a pair of quarks (making it part of a baryon). In either case, we can consider the active quark to be isolated in space.

From the uncertainty principle and the discussion of virtual particles in Chapter III, we know that on a very short time scale it is possible to have processes occur that seem to violate the conservation of energy. One such process would be the creation of a quark-antiquark pair in the empty space near our isolated quark. Any quantum system can fluctuate to a state where energy is increased by an amount ΔE, provided that it returns to normal in a time Δt, where $\Delta E \,.\, \Delta t > h$. There is therefore no reason why quark-antiquark states should not appear. In fact, there is no way we could tell whether they had appeared or not. In this way, the pair is very similar to the virtual particles responsible for the strong force. It is further possible that the isolated quark will combine with the virtual antiquark from the pair, forming a meson, and that the virtual quark will take the place of the original quark in the particle (see Illus. 67). If the original quark happens to be in a meson, such as the ρ^+, then we could have a process similar to that shown in Illustration 68. We see that the original ρ^+ is transformed into two particles with zero spin and baryon number. One of these is positively charged and the other neutral, so it is quite evident that this process describes the decay of the ρ^+ into two pi-mesons.

Actually, this same process of meson creation by the combining of original quarks with quark-antiquark pairs is the mechanism for the

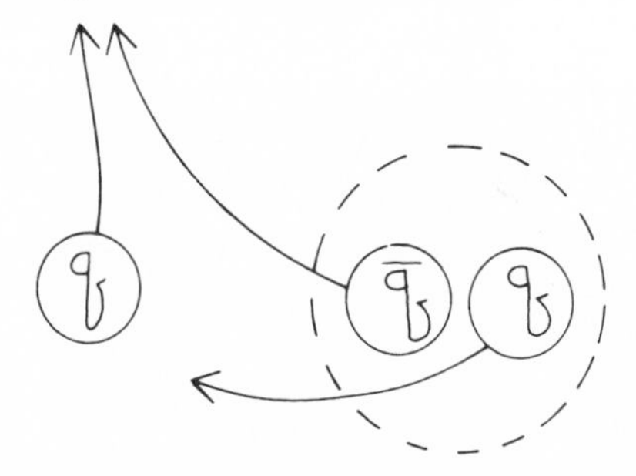

67. Formation of a meson.

baryon decays we talked about in the previous section. What we described as "quark spin flip with pion emission" is actually a quark spin flip accompanied by the recombination of a quark-antiquark pair, which is the same process that occurs in rho decay.

Once we understand the picture of meson decay suggested by the quark model, we can see that all of the systematics of the decay process discussed for baryons can also be applied to the mesons. Mesons containing the strange quark will decay slowly if that quark has to change into a *u* or *d*, but will decay quickly if it does not have to change. As with the baryons, there should be a large number (in principle, an infinite number) of higher mass mesons corresponding to situations where the quark and antiquark circle each other in expanding orbits. For example, the A_2 meson, which has a mass of 1,310 MeV and $J = 2$, would be a state something like that shown in Illustration 69.

Finally, the fact that the decay process is identical in baryons and mesons suggests that it ought to be possible to find relationships between decay rates for mesons and baryons. After all, the process by which a quark combines with a pair should not depend very much on whether the original quark is attached to an antiquark (as it would be in a meson) or to two other quarks (as it would be in a baryon). Thus, even at the theoretical level we see that the distinction between differ-

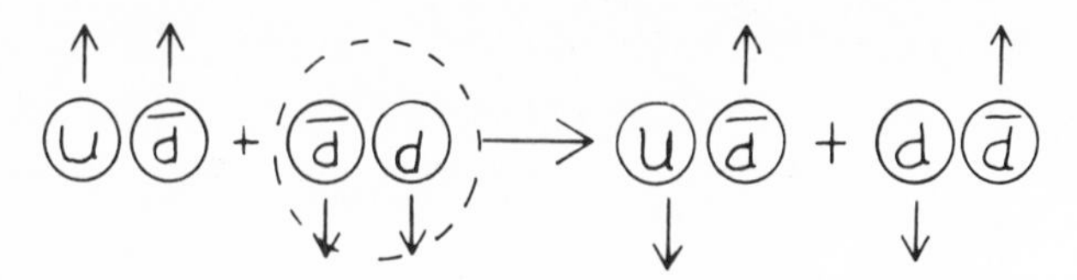

68. The decay of a rho-meson.

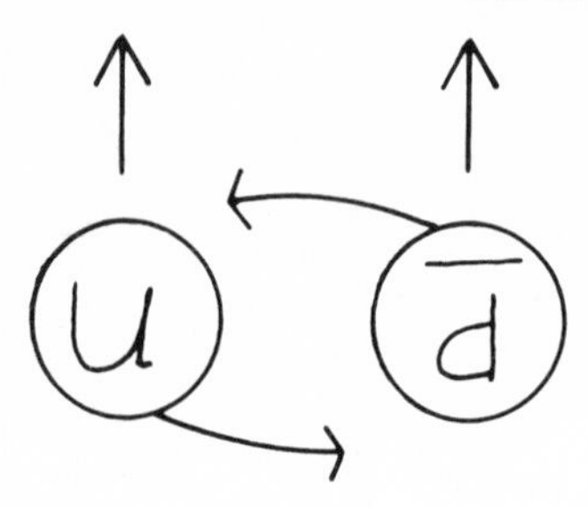

69. The A_2 meson.

ent particles begins to become less important, and the essential fact is that every hadron we have discovered is composed of the same kinds of constituents.

Philosophical Consequences of the Quark Model

THE question we posed at the beginning of the book was this: "Is there an underlying simplicity in nature?" We have now followed the trail of the answer through three periods when the answer seemed to be yes. First, there was the atomic picture of the chemical elements, in which the tremendous variety of materials was perceived to be composed of a relatively small number of atoms. This simplification was followed by a period of increasing complexity, during which the number of known chemical elements swiftly increased. The periodic table of the elements provided a measure of order among the elements, but why it worked was not understood.

The next major simplification was the development of the modern nuclear atom, in which all chemical elements were believed to be composed of three basic building blocks—the proton, neutron, and electron. When this picture was investigated more closely, however, a new kind of complexity developed, a development that we have been following in this book. By the early 1960s, the three elementary particles of the nuclear atom had been superseded by hundreds of elementary particles with various new properties. Once more our picture of the world had become complex.

The quark model therefore represents the most recent step in this dialectic of complexity leading to simplification leading to complexity again. The hundreds of elementary particles are now replaced by three quarks and their antiparticles, and all the particles are thought of as combinations of these three basic building blocks. With the development of the quark model, the answer to our basic question seems to be a resounding yes. Whether we would still answer that way in the future remains to be seen.

Whether it is a permanent simplification or not, the quark model is a tremendous improvement over the previous chaos. It suggests a way of picturing the elementary particles that has been extremely productive in research and that lends itself easily to teaching. I can recall seminars on the quark model where the speakers made their point by employing colored wooden blocks with arrows painted on them to illustrate hadronic structure, which could hardly have been done using something as abstract as the eightfold way.

CHAPTER X

Evidence for the Quark Model

Perry Mason turned. "Circumstantial evidence is the best evidence there is, Paul. You just have to interpret it properly."

—ERLE STANLEY GARDNER,
The Case of the Queenly Contestant

The Quark Model and Particle States

ONE of the major arguments in favor of the quark model is its enormous success in explaining and codifying the hadrons. If we think of the hadrons as particles, all of the simplifications and classifications we have discussed do not reduce the confusion to the point where a pattern can be discerned. Once we have the concept that each of these particles corresponds to a different arrangement of the quarks, however, the situation changes. We can see that the families of particles (such as those associated with the delta and the nucleon, for example) are indeed related to each other in a fundamental way: They are all composed of the same quarks. The patterns of decay speeds associated with strangeness are also understood in terms of the number of *s* quarks present before and after a decay.

Physicists have, for two reasons, spent a lot of time over the past 15 years looking at all of the ramifications of the quark model. One is to see whether any of the predictions of the model are wrong; the other is to see what new information about the particles can be gained.

To understand how the first type of investigation is carried out, we can consider the nucleon. In the quark model, the nucleon is a composite system made up of three particles, each with spin and isospin 1/2, which is not an unfamiliar system in physics. There are nuclei whose constituents are nucleons (which also have spin and isospin 1/2) that are very similar. Tritium (chemical symbol ^3H), an isotope of hydrogen with a proton and two neutrons, and ^3He, an isotope of helium with two protons and a neutron, are both very similar to the quark picture of the nucleon. This similarity was exploited by theoretical physicists to predict what sort of particles one would expect to find in the nucleon family. Calculating the various combinations and permutations of the quarks is not, after all, very different from calculating the permutations and combinations of protons and neutrons that would make up the excited states of ^3H and ^3He. When this calculation is made, an amazing fact emerges: *There is no particle predicted by the quark model that has not been found.* Perhaps even more surprising is the converse statement: *No particle has been found that does not fit into the quark picture.*

Here is probably the main reason why physicists today so readily accept the quark model. We can get some of the flavor of the process of prediction and discovery that has gone on since the early 1960s if we look at the first (and probably best known) particle discovery that has resulted from this process—the particle called the Ω^- (omega minus).

Although the particle was first predicted from the eightfold way symmetry, it is probably easier to understand the prediction in terms of quarks. We have seen that there are particles with no strange quarks, particles with one strange quark, and particles with two strange quarks. These correspond to strangeness 0, -1, and -2-type particles, respectively. Is it not possible that there might be a baryon made up entirely of strange quarks? What would the characteristics of such a particle be?

From the quark table in Chapter IX, we see that such a particle would have a charge

$$Q = -\frac{1}{3} - \frac{1}{3} - \frac{1}{3} = -1$$

and a strangeness

$$S = -1 - 1 - 1 = -3$$

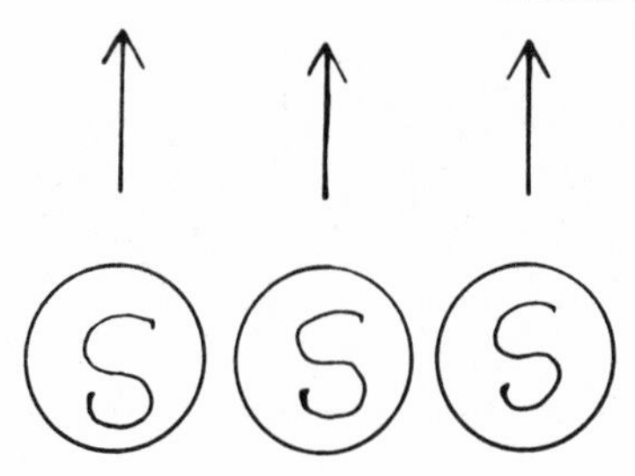

70. The Ω^- particle.

along with the usual baryon number of 1. It would also have its spins aligned, so that it would look something like Illustration 70, above. The mass of the particle was estimated to be about 1,680 MeV, and it was christened the Ω^- (omega minus) even before it was discovered.

Once the prediction was made, the race was on to find the new particle. In February 1964, a group at Brookhaven Laboratory reported the first production of the Ω^- in a liquid-hydrogen bubble chamber. It is characteristic of the scale of modern high-energy physics that there were no fewer then thirty-three persons listed as authors of the paper.

The event they saw occurred when a K^--meson entered the chamber, striking a proton target. The following string of events was seen in the chamber:

$$\begin{array}{l} K^-P \rightarrow \Omega^- + K^+ + K^\circ \\ \qquad\quad \hookrightarrow \Xi^\circ + \pi^- \\ \qquad\qquad\quad \hookrightarrow \Lambda^\circ + \pi^\circ \\ \qquad\qquad\qquad\quad \hookrightarrow p + \pi^- \end{array}$$

Each of the three decays was slow, allowing it to be seen in the bubble chamber. We would expect three slow decays for the Ω^-, of course, since there are three strange quarks to be converted to u and d quarks.

This exciting discovery meant that the new simplicity in particle physics was borne out by a crucial experimental test. I was a graduate student at Stanford at the time; I can remember how rapidly the news spread through the department, with groups of professors and students standing around trying to figure out all that the discovery implied.

As I have said, the original prediction of the existence of the Ω^- was made by using the eightfold way, rather than directly from the quark

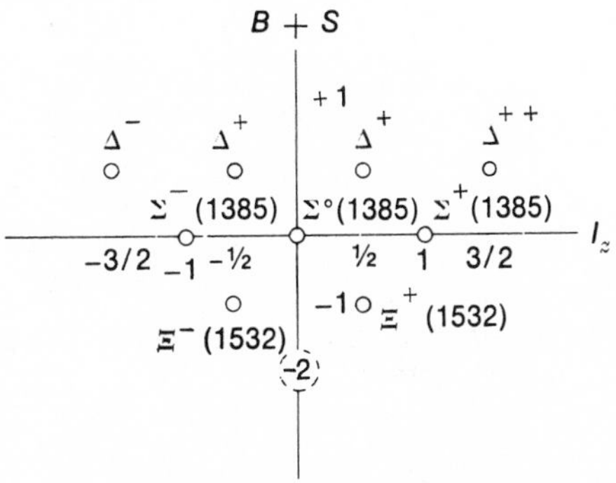

$B + S$

$+1$

Δ^- Δ^+ Δ^+ Δ^{++}

$\Sigma^-(1385)$ $\Sigma^\circ(1385)$ $\Sigma^+(1385)$

I_z

$-3/2$ -1 $-½$ $½$ 1 $3/2$

-1 $\Xi^-(1532)$ $\Xi^+(1532)$

-2

71. The eightfold way grouping of the spin 3/2 baryons.

model. For the sake of completeness, here is how that original prediction was made.

The eightfold way predicts that the low-mass, low-spin particles will group themselves in octets. For higher spins, it predicts that other groupings may occur as well. Consider, for example, the following spin 3/2 particles, which are members of the delta, sigma, and cascade families: $\Delta(1236)$, $\Sigma(1385)$, and $\Xi(1532)$. The last particle in this grouping has not been introduced before, but it is one of the members of the cascade family that we mentioned briefly in Chapter IX. If we make an eightfold way plot of these particles, a plot in which each particle represents a point on a chart whose axes are $(B + S)$ and I_z, we will get the graph depicted in Illustration 71. The mathematics predicts that there should be ten particles in this grouping, and it does not take much imagination to see where the missing particle should go. From the graph we see that it should be a particle with $B + S = -2$ and $I_z = 0$, which means that if it has $B = 1$, it must have $S = -3$ and $Q = -1$, which is, of course, what we require for the Ω^-.

If we consider the prediction for a moment, we can see an analogy with the development of the periodic table of the elements. There were "holes" in the table that led to the discovery of new elements, and in the same way, a "hole" in the elementary particle table led to the discovery of the Ω^-.

We can even understand how the prediction of the mass of the new particle could be made. A little arithmetic on the masses of the particles listed above reveals the following facts:

$$M_{\Sigma(1385)} - M_{\Delta} = 1{,}385 - 1{,}236 = 149 \text{ MeV}$$

and

$$M_{\Xi(1532)} - M_{\Sigma(1385)} = 1{,}532 - 1{,}385 = 147 \text{ MeV}$$

In other words, the difference in masses between the members of the grouping is roughly 150 MeV. It would be reasonable to suppose the Ω^- would have a mass about 150 MeV higher than the $\Xi(1532)$. This, in turn, leads to a predicted Ω^- mass of

$$M_{\Omega^-} \approx 1{,}680 \text{ MeV}$$

The modern value is 1,672 MeV—a remarkable agreement.

Thus, both the qualitative and quantitative predictions of the quark-model–eightfold-way were verified by the discovery of the Ω^-. Since 1964 physicists have by and large regarded this discovery as one of the most important pieces of evidence for the existence of quarks.

The Quark Model and Scattering

WE have seen how important new information about elementary particles can be gained by measuring the probability that two particles will interact, or, as discussed in Chapter VII, the interaction cross section. It was measurements of this type that led to the discovery of the delta and, ultimately, of the whole family of nucleon and delta resonances. A large body of data on scattering cross sections exists, and it should not be a surprise that the quark model can be used to deal with it.

To understand how this can be done, we can consider the scattering of two baryons, which is a situation we could produce by directing a proton beam at a hydrogen target. From the point of view of the quarks, this would correspond to one object made of three quarks scattering from another object that was also made of three quarks. One way to imagine the scattering is to think of one quark from the projectile scattering from one quark in the target and both moving off, dragging their respective partners with them. (See Illus.

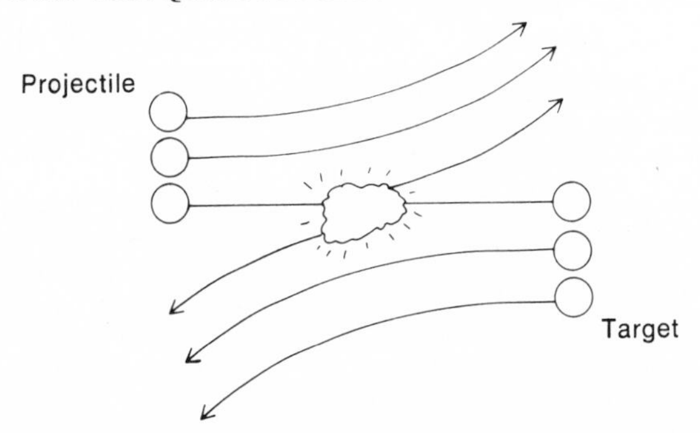

72. The scattering of two baryons.

72.) In proton–proton scattering there are precisely nine ways in which this kind of interaction can happen, corresponding to each of the three quarks in the projectile scattering from each of the three quarks in the target.

The cross section for scattering, you will recall, is related to the probability that a reaction will take place. The picture we have just described relates the probability that a proton will interact with another proton to the probability that a quark will interact with another quark. If we denote the cross section for one quark to scatter from another by σ_q, then in this simple picture of the scattering process, the proton–proton scattering cross section must just be $\sigma_{pp} = 9\sigma_q$.

So far, all we have done is to replace one number we are not able to calculate, σ_{pp}, by another number we are not able to calculate, σ_q. But we can go farther if we consider the situation pictured in Illustration 73 (on page 156), which is the scattering of a meson from a proton, as conceived in the quark model. The meson is made up of two quarks (we ignore for the moment the differences between quarks and antiquarks), and each of these can scatter from each of the three quarks in the target proton. An argument similar to the one given previously tells us that the meson–proton cross section must be $\sigma_{Mp} = 6\sigma_q$, where again σ_q characterizes the scattering of one quark from another.

At first glance, it would seem that we have just obtained another relation between two quantities that we cannot calculate. But, if we divide the equation for σ_{pp} by the equation for σ_{Mp}, we get

$$\frac{\sigma_{pp}}{\sigma_{Mp}} = \frac{9\sigma_q}{6\sigma_q} = \frac{3}{2} = 1.5$$

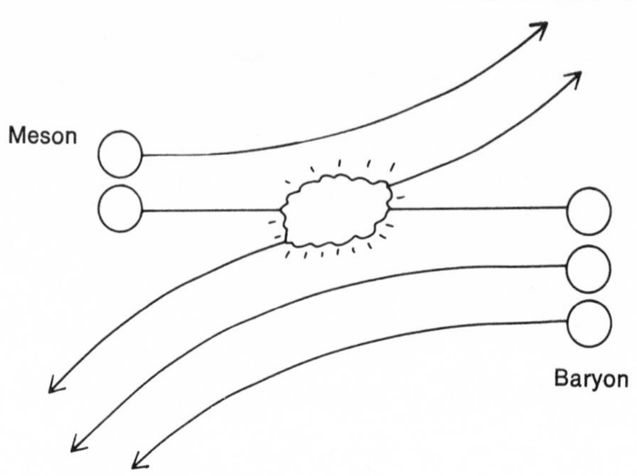

73. Scattering of a meson from a proton.

and this does *not* depend on anything having to do with the quarks. It is, in fact, a clean prediction that tells us that proton–proton cross sections ought to be half again as big as meson–proton cross sections.

A typical value of σ_{pp} might be 45×10^{-27} square centimeter, while a typical value of σ_{Mp} might be 27×10^{-27} square centimeter. For these numbers, the experimental ratio is

$$\frac{\sigma_{pp}}{\sigma_{Mp}} = \frac{45}{27} = 1.65$$

which is close enough to 1.5 for us to say that the prediction of the model has been verified.

This is a good example of the way one can work with the quark model. The strategy is to calculate measurable quantities (like σ_{pp} and σ_{Mp}) in terms of quantities that relate to the quarks and then to do enough algebra to eliminate the latter quantities from the equations, leaving equations that link measurable quantities. These relations can then be tested against experiment, as above.

During the middle and late 1960s many relations between quantities involved in particle scattering and production were derived by using the quark model. At the level of 20 percent accuracy, virtually all of them turned out to be correct. This is another strong indication that particles are composed of quarks.

A Rutherford Experiment for Quarks

ONE of the strongest pieces of evidence for the existence of the atomic nucleus was the experiment described in Chapter I. From the way alpha particles scattered from atoms, Ernest Rutherford was able to conclude that most of the mass of the atom was concentrated in a small nucleus. In 1969 a collaboration between physicists at the Massachusetts Institute of Technology and the Stanford Linear Accelerator Center (SLAC) seemed to produce a similar result for the proton. Their results suggest that the proton itself is made up of small constituents.

We have already given a brief description of the Stanford accelerator. It accelerates electrons down a 2-mile tube, bringing them to energies of over 20 GeV by the end of the trip. In an area called the beam switchyard, a series of magnetic selectors and slits are used to produce an intense beam of electrons whose momentum is very accurately known. These electrons are then allowed to strike a target in the experimental areas beyond the switchyard.

In the M.I.T./SLAC experiment, the target that the beam encountered was liquid hydrogen, so the basic interaction being studied was between electrons and protons. After the impact, the direction and momentum of those electrons that had suffered collisions were measured. In Illustration 74, below, we show this experiment schematically and, to emphasize the logical similarity with Rutherford's experiment, we include a similar sketch of his method. In each experiment a projectile is directed against the target whose properties we wish to study. The idea is that by looking at what happens to the projectile after it hits the target, we can learn something about the way the target is con-

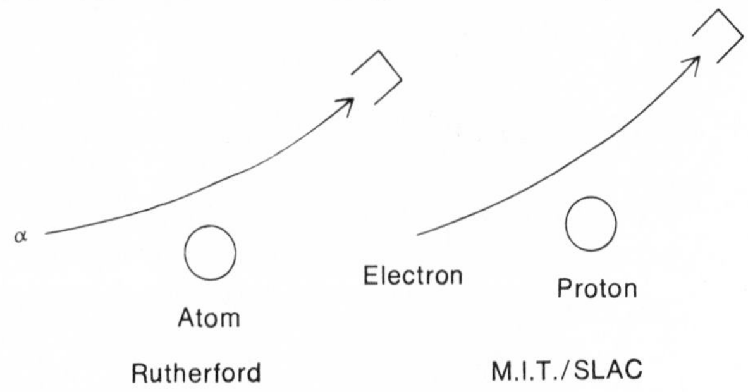

74. Similarities between the M.I.T./SLAC experiment and Rutherford's experiment.

structed. This certainly succeeded in the Rutherford experiment, where the large number of particles coming off backward was a direct indication of the existence of the nucleus.

Of course, there are also differences between the two experiments. Rutherford's alpha particles had energies of only a few MeV, so they could not create any new particles in the collisions. The electrons, having in excess of 20 GeV, can (and do) create all sorts of particles when they collide with the proton. The debris of the collision is not measured, but by measuring the momentum and energy of the electron before and after the collision, we can determine how much energy was delivered to the proton and, hence, what the total energy of the final state of the target must be. For example, if the electron gives up 5 GeV in the collision and the end product is a proton and three pi-mesons, we can say with certainty that the sum of the energies of the four particles must be 5 GeV (plus, of course, the mass energy of the proton, which is the same before and after the collision).

A second important difference is the size of the object we believe we can "see" with the projectile in the two experiments. We can get some rough idea of this quantity from the uncertainty principle for momentum and position: $\Delta x \, . \, \Delta p \leq h$. If we use the momentum of the particles as an estimate for the certainty in momentum, and identify Δx as the size, D, of an object we could "see" with the particle, we see that D is just

$$D \cong \frac{h}{p}$$

In the following table we give D for some sample projectiles:

PROJECTILE	TYPICAL MOMENTUM	D
	(eV/c)	(cm)
Photon	10	5×10^{-6}
Alpha	4×10^{6}	10^{-11}
SLAC electron	10^{10}	5×10^{-15}

We infer that a normal photon can "see" something approximately the size of an atom, a Rutherford-type alpha can "see" something approximately the size of a large nucleus, and an electron from SLAC can "see" something much smaller than the size of a proton. Thus, if any experiment has any chance of "seeing" the quarks inside the proton, this one should.

Before the experiment was performed, theorists did not expect to see many scattered electrons with final energies around 8–10 GeV (indicating energy transfers to the proton of many GeV). When the experiment was completed, however, an appreciable number of these electrons was seen—the number exceeding even the most optimistic theoretical estimate by a factor of 40. And, as in the case of Rutherford, these excess particles were seen at large angles.

This result can be understood in analogy with the Rutherford experiment if we think of the electron as scattering off tiny, pointlike particles inside the proton. Thus, the Stanford experiment provided the first direct evidence that elementary particles are not elementary, but are composed of constituents.

Since that time, a large number of other experiments of this general type have been performed. They are called deep inelastic scattering experiments because they probe to very small distances in the proton (and, hence, to "deep" beneath its surface) and create many particles in collisions. The general result of all these experiments seems to be the same: When you work with particle structure at small distances, the particles appear to be made up of constituents, even if you cannot detect these constituents directly in the debris of the interaction.

The Fly in the Ointment

WITH all this evidence in favor of the quark model, it is not hard to understand why it is popular with physicists today. Nevertheless, there is an uneasy feeling about the whole situation. If everything is made of quarks, why do we not see any quarks in the laboratory when we let high-energy hadrons collide? Remember that soon after Rutherford discovered the nucleus, physicists began to see nuclei and pieces of nuclei being created and identified. The seeming reluctance of the quarks to go through the same process remains one of the major difficulties with the model—a difficulty we will examine in detail in Chapter XI.

CHAPTER XI

Where Are They?

We seek him here, we seek him there,
Those Frenchies seek him everywhere.
Is he in heaven?—Is he in hell?
That demmed, elusive Pimpernel?

—BARONESS ORCZY,
The Scarlet Pimpernel

Introduction

SINCE the suggestion in 1964 that quarks might exist, a vast amount of effort to "bring one back alive" has been expended by experimental physicists. As I write in spring 1979, there is only one experimental claim for the discovery of a quark that has not been rejected by the physics community, and that candidate is still awaiting confirmation by independent experimenters. It is fair to say, therefore, that 15 years of searching have failed to produce a single particle in the laboratory that is generally accepted as a quark. A review of some of these searches and a discussion of the response that theoretical physicists have made to this disappointing result are the subjects of this chapter.

The property of quarks that would make them easy to find is, of course, their fractional electrical charge. The standard detection instruments for particles—the Geiger counter, cloud chamber, and bubble chamber—all depend, ultimately, on the electrical interaction of the particle with stationary atoms. In the cloud chamber, for example, the charged particle passing an atom creates an ion that then serves as a nucleus for the condensation of a droplet. In other devices, the ion is detected in different ways, but all the detectors start with the detected particle removing an electron from an atom.

The ability of a particle to create ions depends on the particle's charge. This is as we might expect—the bigger the charge of the particle, the greater the force exerted on the atomic electron. If the charge of the particle is denoted by Q, then it turns out that the number of ions that charge will produce along each centimeter of its track will be proportional to Q^2. This means, for example, that an alpha particle with two protonic charges will create four times as many ions as a single proton or electron moving at the same velocity. We will see a much more dense collection of droplets for the alpha than we would for the proton, and the droplet density can be used as a way of identifying the alpha and distinguishing it from singly charged particles.

For quarks, with their fractional charge, the same thing can be done, except that now we have to look for particles producing *fewer* droplets than would a proton or electron. A quark of charge 2/3, for example, should produce a track whose droplet density is $(2/3)^2 = 4/9 \approx 1/2$ of the density caused by a normal particle, while a quark of charge 1/3 should produce a track of density 1/9. Hence, if we were looking for quarks in a cloud or bubble chamber, we would look for tracks on which we counted too few droplets or bubbles. These are called *lightly ionizing* tracks, and they form the basis for many of the quark searches that have been done in cosmic rays and on accelerators.

An alternate way of utilizing the quark's fractional charge in experiments is based on what happens when a quark is absorbed in matter. There are only two places a quark can go in an atom. It can either enter the nucleus or it can go into orbit and replace one of the electrons. In either case, the resulting "quarked atom" will have a net electrical charge. This means that we can extract the atoms containing quarks by passing the material between plates that are electrically charged. The quarked atoms will tend to collect on the plates while the normal atoms, being neutral, will be largely unaffected. The collected material is presumably rich in quarks and can then be analyzed in more detail.

A large number of searches have been made using this technique. I refer to them generically as *geological* searches, since most of them involve analyzing kinds of materials which, for one reason or another, are thought to be a likely final repository for quarks—moon rocks, sea water, and even oyster shells. The term geological is a little misleading, but it is convenient.

The fractional charge of the quarks inspires this type of experiment because once a quark has been created, it cannot decay into ordinary particles. Such a decay would necessarily violate the principle of charge conservation. Hence, it would seem that once a quarked atom is formed, it will persist and be at our disposal when we start looking for it.

Unlike the direct searches, a negative result in a geological search is a little difficult to interpret. Not finding a quark in sea water, for example, might mean that there are no free quarks, or it might mean that one of the assumptions in the chain of reasoning that led to choosing sea water for the search was wrong. This should be kept in mind as we move on to other quark searches.

Direct Quark Searches

By far the most straightforward way to look for a quark is to repeat the procedures that led to the discovery of so many of the other particles. We could, for example, see if quarks can be found in cosmic ray experiments, either in showers or in the primary cosmic radiation itself. Alternatively, we can try to create quarks in an accelerator and see if we can detect them.

As far as cosmic ray experiments go, there are two possibilities. Either quarks are themselves present in the primary cosmic radiation that falls on the earth, or they are created in the very high-energy collisions that primary cosmic rays make when they enter the atmosphere. To test the first hypothesis, it is only necessary to put out counters and look for lightly ionizing tracks. Because quarks lose less energy through ionization than normal particles, they ought to penetrate farther into the atmosphere and, hence, be more visible in such experiments. In any case, over twenty such searches have been carried out without success, and taken together these experiments can be used to establish an upper limit on the number of quarks falling on the surface of the earth. If we call this limit N, then the number of quarks that actually fall on the earth must be less than N. From the experiments, $N \approx 10^{-10}$ quarks/centimeter/second.

To get some idea of what this limit means, we note that there are roughly 3×10^7 seconds in a year. Thus, if we picked 1 square centimeter of the earth's surface and waited for a quark to hit it, we would have $10^{-10} \times 3 \times 10^7 = 3 \times 10^{-3}$ quarks/year. This means that on the average we would have to wait about 330 years before we could expect to see a quark fall on our target. And, since this represents an upper limit, we could easily have fewer (or even zero) quarks falling, corresponding to a waiting time of hundreds, or even an infinite number, of years. Clearly, if there are quarks around they are not copiously represented in the primary cosmic ray flux.

If this is true, we can still borrow a leaf from history and look for quarks in the debris of cosmic ray collisions. That is how the positron and the mesons were first seen. This type of experiment is usually done by setting out small detectors—detectors that will be triggered when shower particles begin arriving at a site. This signal is then used to trigger the main detector (for example, a cloud chamber would be expanded only when the small counters indicated that a shower was present). In this way, the quark search would only take place when a major cosmic ray event had occurred. In a variation on this device, some experiments have scanned particles arriving after the shower, in case the quarks should be slow and heavy.

Experiments of this type have been carried out in many places. For a time in 1969 there was a flurry of excitement when a group in Australia reported a couple of events that might have been caused by quarks, but this quickly died down when an improved experiment by that group and duplicate experiments elsewhere failed to turn up any similar events. In general, the shower experiments give about the same limit as the single-particle experiments—a quark flux at the earth's surface of less than 10^{-10} quarks/second/square centimeter.

The failure of cosmic ray experiments caused physicists to turn to accelerators. In fact, there is a ritual search made for quarks whenever a new high-energy machine is turned on. These searches have the advantage of being able to control the incident beam of projectiles and the disadvantage of being limited in energy by the machine design. Quarks must be produced in pairs in order for charge to be conserved, which means that there is a limit to the mass of a quark that can be produced in any accelerator, a limit given by the mass-energy relation discussed in Chapter IV. Typically, this limit will be from a few GeV to 15–20 GeV, depending on the machine. We now look at a typical accelerator search so that we have some idea of how they are done.

In Chapter VI we saw that the radius of a circle through which a magnetic field B will bend a particle with momentum P is given by

$$R = \frac{P}{Bq}$$

where q is the charge of the particle. In Chapter VII it was explained how this fact could be used to make a magnetic spectrometer. By a slightly different chain of reasoning, it can also be used to turn such a spectrometer into a highly efficient quark detector.

Consider the situation shown in Illustration 75, below: A beam of protons strikes a target, and the collision products go through a magnetic analyzer. For the sake of definiteness, let the incident proton have a momentum P_p. Then the highest momentum that any particle coming out of the interaction can have is P_p (otherwise, momentum could not be conserved). If such a particle has a normal unit of charge, it will be bent through some radius by the magnet. Call this radius R_n, where the n denotes the radius through which the fastest particle with unit charge is deflected.

Suppose now we have reason to believe that a quark of charge 1/3 and momentum P is produced in the interaction. The radius through which it will be bent (call it R_Q) will be

$$R_Q = \frac{P}{B \cdot \frac{1}{3}e} = 3\,\frac{P}{Be}$$

where e is the charge of the electron. It can, in other words, be larger than R_n, the largest radius through which a normal particle can be deflected. Consequently, if we build our apparatus as shown, with the

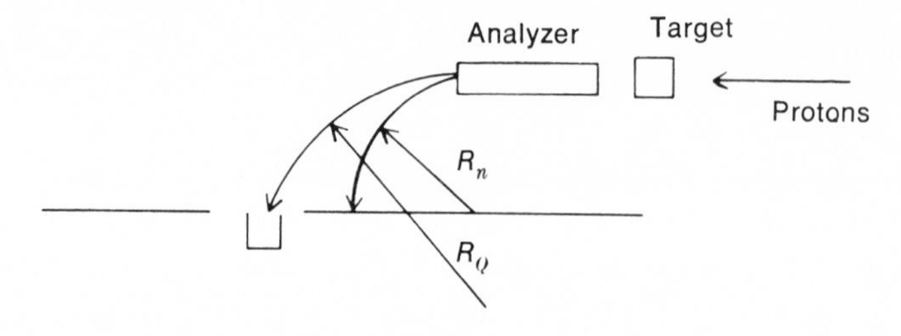

75. The spectrometer as quark detector.

detectors behind a slit set for a radius R_Q, momentum conservation tells us that no particle of normal charge will ever enter the detector. Consequently, if anything is seen, it must have less than a unit charge, and some simple ionization measurements will tell us immediately if it is a quark.

Thus, by using our knowledge of the energy of the incident beam, it is possible to design very clean experiments for quark searches. When these experiments are carried out (and dozens have been), no quarks are seen. This fact is then used to put an upper limit on the probability that a quark is produced in a proton–proton collision. The best limits now suggest that a quark will be produced in less than one interaction in a trillion (10^{12}), with a strong possibility that it will not be produced at all.

Geological Searches

THE logic for geological quark searches is somewhat different from that of the direct type of measurements discussed in the last section. We can start by assuming that quarks are present in cosmic rays, but at a level just below the limits set by experiment. Assume, in other words, that quarks are striking the earth at the rate of about 10^{-10}/square centimeter/second. The direct searches are not sensitive enough to detect this small number of quarks, and it is the function of geological searches to remedy this defect.

If quarks have been falling on the surface of the earth at this rate since the earth was formed 5 billion years ago, then roughly 10^8 quarks would have fallen on each square centimeter of the earth's surface. These quarks have to go somewhere and, since they cannot decay, they must still be on the earth. The objective then becomes one of trying to guess where they are right now.

If we think about the question, we realize immediately that we have to make some assumptions about long-term geological shifts in the earth's surface. A rock that sits on the surface today, exposed to cosmic rays, may have been deep underground a million years ago, and may be deep underground a million years from now. Hence, any quarks that strike the surface will become mixed into the ground to some depth, and what we assume about that depth makes a great deal of difference as to the number of quarks we expect to find per cubic centimeter in geological material. For example, if we assume the depth of the mixing is a few kilometers, we would expect about 1,000 quarks per cubic centimeter in the earth's surface. If we assumed the mixing was more or less uniform, there would be about that number of quarks in any randomly selected piece of material. The number corresponds to about one quarked atom for every 10^{21} normal atoms, a number that is just barely detectable by the techniques we will describe in this section.

But consider the long chain of reasoning involved in this conclusion. First, we have to assume that there actually *are* quarks hitting the earth. Then we have to assume that these quarks are incorporated uniformly into surface atoms. Then we have to assume something about the geological processes that carry these atoms around. A negative result could arise from any one of these assumptions being wrong. This is what I meant earlier when I said that geological quark searches are harder to interpret than direct ones.

Most people who have carried out geological searches have not assumed a uniform distribution of quarks in the earth's materials, but have reasoned from the properties of quarks to estimate their concentration. For example, in 1968 David Rank at the University of Michigan, while working with the theory for "quarkium" (an atom in which a single electron circles a quark), concluded that this particle ought to be soluble in water and behave somewhat like lithium. From this inference he concluded that quarks ought to congregate in the sea, where biological processes might concentrate them further. Consequently, he analyzed sea water, seaweed, oyster shells, and plankton to see if he could find any quarks. Since this method is rather typical of geological searches, it is worth examining in a little more detail.

First, the various materials were ground up and heated until they vaporized. The vapor was passed through an electric field so that any charged atoms were concentrated on the charged plates (see Illus. 76). This concentrated material was then tested by heating it again in an electric arc and analyzing the light given off. Electrons around a quarked nucleus will have different orbits from those in a normal atom because the nucleus has a different charge. Therefore, from the discussion in Chapter I, we know that such atoms should emit different photons, and these photons, if detected, would be one kind of evidence that we could use to deduce the presence of quarks in the sample. From this sort of analysis, Rank concluded that if there were any quarks in his

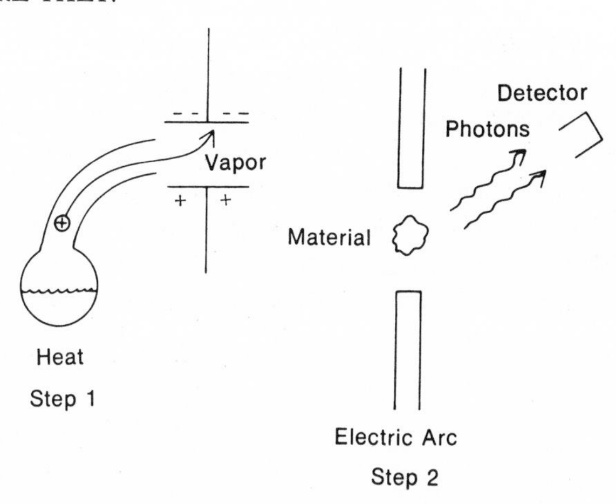

76. Diagram of apparatus used in quark search.

sample, there were fewer than one for each 10^{18} atoms of sea water, and fewer than one for each 10^{17} atoms in the seaweed, oysters, and plankton.

Another common method of analyzing concentrated materials is called the oil drop experiment. This is an old technique that was first used in 1910 by Robert A. Millikan to measure the charge of the electron. In Rank's work at Michigan, various organic oils (such as peanut and cod liver) in drop form were tested directly for quarks. It would also be possible to make a sample by dissolving some concentrated material in oil. In any case, the oil is sprayed into the air between plates whose voltage can be regulated (see Illus. 77). There are then two forces acting on the drop: gravity (tending to pull it down) and electricity (tending to lift it up). The electrical force depends on the charge of the drop. By

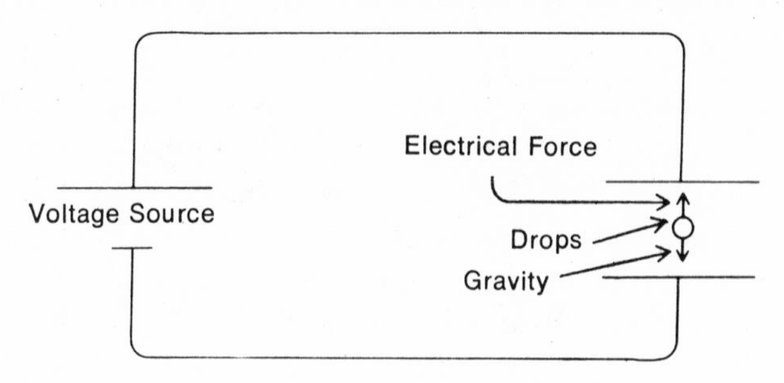

77. Diagram of apparatus used for the oil drop experiment.

adjusting the voltage on the plates and watching the drops move up and down, it is possible to determine this charge with very high accuracy; certainly well enough to tell whether it is fractional or not.

Rank was able to conclude that if quarks existed in his oil samples, there was less than one for each 10^{19} ordinary atoms. In the following table we give a sampling of the materials that have been tested and the number of normal atoms in which a single quark would have to be diluted to explain the negative result of the experiment.

MATERIAL	NUMBER OF NORMAL ATOMS (rounded off to nearest factor of 10)
Lava	10^{22}
Grand Canyon rock	10^{22}
Sea water*	10^{25}
Ocean sediment	10^{21}
Meteorites	10^{17}
Iron	10^{21}
Moon rocks	10^{22}

*This is a more recent experiment than the one discussed in the text.

Unfortunately, in this short section we cannot do justice to all the clever techniques that have been devised to find geological quarks. But with the exception of the experiment we are about to discuss, all of them have produced limits on the quark concentration similar to the ones in the table.

The Stanford Experiment

THERE is a special limbo in science for controversial experiments that cannot be proved wrong, but which have not yet been duplicated by independent work. In 1977 scientists working in the laboratory of William Fairbank at Stanford University reported that they had seen fractionally charged matter in an experiment. This is the only claim for quark discovery that has not been largely rejected by the scientific community, but it would be wrong to conclude from this that the claim is now accepted. The present attitude is one of "wait and see." If others can reproduce the Stanford result, it will be acknowledged as the first discovery of the quark. If it cannot be reproduced, it will probably not gain widespread acceptance.

The experiment is an updated version of the oil drop just described. Instead of an oil drop, a niobium pellet, cooled to a temperature close to absolute zero, was placed in a magnetic field. At this temperature, niobium has the property of being superconducting—a property which, for our purposes, implies that electrical currents can flow in the metal without any concommitant energy loss. It turns out that in this situation, the currents on the surface of the pellet interact with the magnetic field in such a way as to produce a force on the pellet that can oppose the downward force of gravity. The niobium pellet, whose mass is roughly 10^{-4} gram, is literally "levitated" by the magnetic field; it is held up without any contact with its surroundings.

If electrical forces are now applied to the pellet, it will move, and the way it moves will depend on its total charge. Rather than measure the total charge, the Stanford group elected to change the charge of the pellet by irradiating it, a process that creates ions and therefore supplies a means of moving a charge onto or off the sample. This process was repeated until the charge got as close to zero as they could make it. It should be noted that this technique of changing the charge of a sample has been used for years in oil drop experiments, and should not cause any unexpected side effects.

On three of the nine pellets tested, the Stanford group found a residual charge of 1/3 (either plus or minus). In announcing this result, Fairbank used a very useful analogy to describe the technique. Suppose, he said, that you had a bank account, and suppose you could only deposit or withdraw money from the account in whole numbers of dollars. You could start withdrawing and depositing until you got the balance of the account as close to zero as it would go. If you had 25 cents left in the account, that would be very good evidence for the existence of quarters. In the same way, the fact that adding and subtracting whole numbers of charges to the niobium pellets leads to a fractional residual charge is evidence for the existence of fractionally charged particles—the quarks.

Paradoxically, the reaction of the scientific community to this result might have been more favorable if only one quark had been seen, rather than three. The result works out to one quark for each 10^{20} atoms of niobium. If we assume that quarks are uniformly distributed in matter, then this is a clear contradiction of many other geological searches. On the other hand, when dealing with an unknown quantity such as a quark, we should not be too quick to accept such an assumption. It is possible that there is something about the chemistry of quarked atoms that make niobium an ideal material to hold them. It is possible, for example, that *all* of the geological quarks on Earth are in niobium atoms. Not likely, of course, but possible. Hence, we must withhold judgment on the Stanford experiment until we see what the next few years bring us.

What Does It Mean?

THE brutal fact of the matter is that there is not a single generally accepted piece of evidence that a quark has been isolated in a laboratory. What are we to make of this?

One way of dealing with the fact is to advocate the mathematical quark argument. If you look at the evidence for the quark model in Chapter X, you will realize that all that was shown was that elementary particles act *as if* they were made up of quarks. Nowhere was it necessary to assume that quarks had to exist as free particles. It can thus be said that quarks are just a mathematical fiction that happen to make it easy to talk about elementary particles, but need have no more real existence than a degree of temperature.

Perhaps an analogy will help: We know that waves can be seen on water, and that many properties of water can be described in terms of waves. But do waves exist in the sense that we are requiring that quarks exist? If you started dismantling a sample of water, you could go right down to a pile of atoms without finding anything you could point to as a "wave." The concept of a wave is just a useful way of describing something that happens when large numbers of water molecules act together in a certain way. Similarly, it is quite possible that we could take elementary matter apart and not find something we could label "quark," although the matter in bulk could exhibit the properties we have described as evidence for the quark model.

A more modern version of this argument is called the *confinement theory,* which holds that quarks really do exist in the elementary particles and that inside the particles they really behave just as we would expect consituents to behave. But the theory postulates that something prevents these constituents from being taken out of the particle. Such an idea would account for the success of the quark model as well as for the absence of free quarks in the laboratory.

There are a number of ways in which we can imagine quarks being "confined." The simplest way is to assume that the force that binds the quarks together into hadrons is just too strong to be broken by the amounts of energy we are able to supply. For example, we know that it takes several MeV to knock a nucleon out of a nucleus. If we had not known this fact and had tried to find the nucleon by bombarding nuclei with projectiles with energies in the keV range, we might have concluded that even though nuclei behaved as if they were made up of protons and neutrons, these hypothetical particles could not exist in the conventional sense. We might even have spoken of mathematical protons. The possibility always remains that we just have not been able to produce particles of high enough energy to shake a quark loose. This possibility is the reason that quark searches are always one of the first experiments to be done on new accelerators.

But present-day theories of high-energy interactions suggest another, more subtle way in which quarks might be "confined." In these theories, the mathematics leads us to picture the quarks as being bound together in much the same way as the two ends of a rubber band are held together—by the band itself. What we call a quark might be one end of the rubber band. A meson, in this picture, would look something like the sketch shown in Illustration 78 on page 172.

It is evident that in such a situation it is logically impossible to create a quark outside of a particle. If you break the string to free one of the quarks, all you will get are two shorter pieces of string. (See Illus. 79.) Since each of these will have two ends, this process will be interpreted as the production of a meson in the laboratory. In a sense, therefore, the question "Where are the quarks?" has already been answered in this model; that is, the breaking of the string is responsible for the flood of mesons that are created in high-energy interactions. This model clearly encompasses the two aspects of the quark model we have discovered so far: It gives us particles that are made up of quarks, but it prevents the quarks from appearing as free particles.

There are several different versions of this theoretical confinement scheme, and which one (if any) turns out to be right is something for the future to tell. We simply note that the theory of the quark model

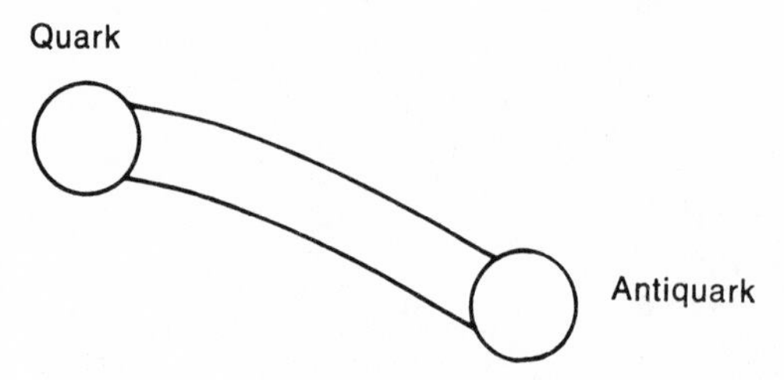

78. Representation of a way quarks could be "confined" in a meson.

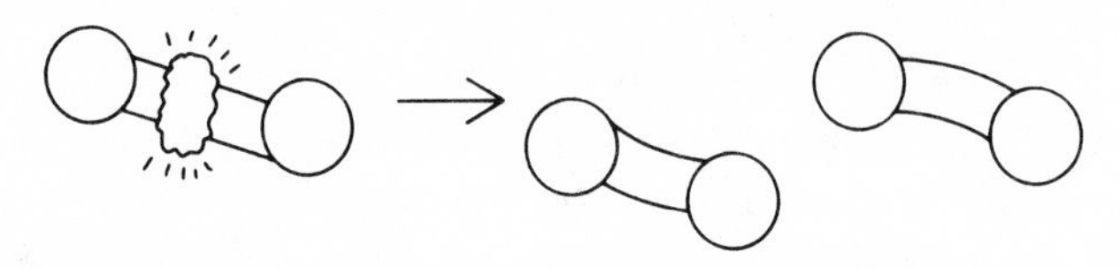

79. Theoretical result of breaking the string to free a quark.

has progressed to the point where even a total failure to find a free quark would not necessarily mean that the model should be abandoned.

Still, the idea that something can exist and yet not be visible in an experiment, even in principle, raises some tricky philosophical questions, to put it mildly. At present, the mathematical and confined quarks are discussed because the data seem to favor these ideas. The whole business was put into perspective by Lawrence Jones of the University of Michigan, who carried out a quark search himself and authored a recent review of the data: "It is," he said, "a question on which wise men can differ."

Enough said.

CHAPTER XII

Charm and the Proliferation of Quarks

Be fruitful, and multiply . . .
—Genesis 9:1

The Discovery of the ψ/J

WE have seen that the strongest motive for using the quark model is the simplicity that it introduces into our picture of elementary particles. By the mid-1970s, however, some developments had taken place that cast serious doubt on the ultimate simplicity of the whole quark picture. One set of developments was experimental and rather unexpected. It involved the discovery of new particles. The other was theoretical and, in a sense, was part of the model from the very beginning. We can begin with the experimental situation.

In mid-1974 two experiments were being conducted several thousand miles apart. The two were completely different and the physicists involved in one had no idea of what was going on in the other. Yet each of these efforts resulted in the discovery of a new particle, and these discoveries were made so close together that they were reported in the same issue of *Physical Review Letters.*

One experiment was being done at Brookhaven National Laboratory on Long Island under the direction of Samuel C. C. Ting of the Massachusetts Institute of Technology. A beam of protons was directed against a target made of beryllium (a lightweight metal whose nucleus contains nine protons and neutrons). A series of magnetic selectors, scintillators, and Čerenkov counters were set up in a two-armed arrangement, as shown in Illustration 80. Its purpose was to look for reactions of the type

$$P + B_e \rightarrow e^+ + e^- + \text{anything}$$

In other words, the M.I.T. experiment was designed to look for electron pairs produced in the proton–nucleus collision.

This is a very difficult experiment to carry out. In Chapter VII we saw how extreme care had to be taken in order to distinguish between rare antiproton production and the copious production of pi-mesons. In Ting's experiment the difficulty is much greater, since electron pairs are produced very seldom, particularly at the large angles where the experiment was designed to make measurements. There could be billions of hadron pairs produced for each electron pair.

To get some idea of what this means, think of a rain shower falling on an average-sized city. There might be several billion raindrops in a typical shower. Finding the electron pair among all of the hadrons is analagous to finding one raindrop falling on a city during a shower—not the easiest of tasks.

On the other hand, Ting and his group had spent over 10 years perfecting the apparatus and, in the process, acquiring a solid reputation as extremely careful and precise workers. I can testify to the validity of this reputation personally, because I spent two summers with the group when it was working in Hamburg, Germany. My most vivid

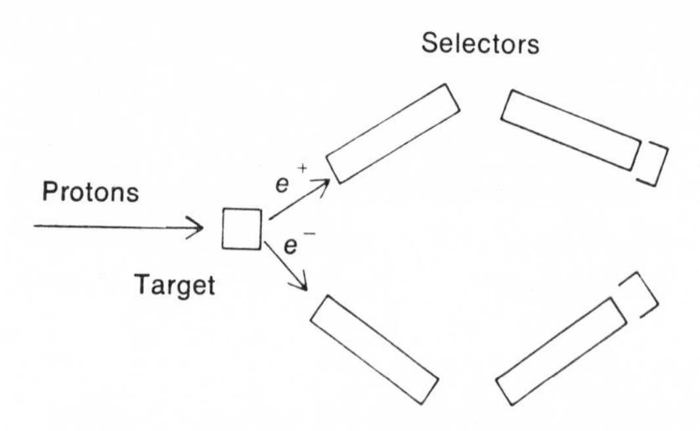

80. Diagram of apparatus used in the experiment when the ψ/J was discovered.

recollection of the "check–double-check–triple-check—and then check again" operation was that each computer program used to analyze data was written independently by two physicists in the group, and the results of the independent programs were then compared at the daily staff meetings. Only when everything checked out would the data analysis go ahead.

Using this careful experimental technique, the group began collecting data in the late summer and early fall of 1974. When they plotted the number of electron-positron pairs as a function of the energy of the pair, they got a result like that pictured in Illustration 81. A large peak (corresponding to about 250 events) appeared at an energy of about 3.1 GeV. Recalling our discussion of phase space diagrams in Chapter VII, we realize immediately that this must mean that the reaction must proceed via the production of an intermediate particle, as shown in Illustration 82.

The new particle was christened the "J" by the M.I.T. group. There are various reasons given for this name. The official reason given by

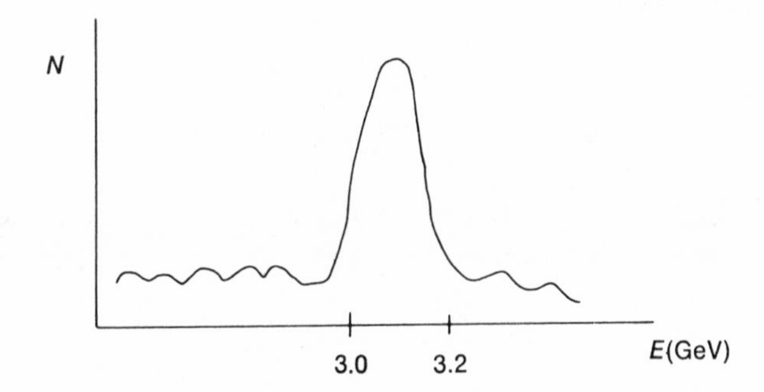

81. Graph of electron-positron pairs.

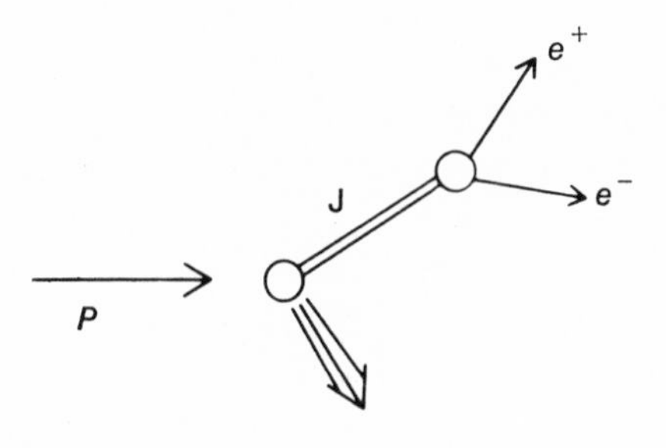

82. The production of the "J" particle.

Ting is that the particle resulted from an interaction involving the electromagnetic current, usually denoted in theoretical physics by the letter "J." Others have pointed out, however, that there is a striking resemblance between the letter "J" and the Chinese character for Ting. In either case, it was clear that the group had discovered a new particle.

While this was happening at Brookhaven, another group at Stanford, led by Burton Richter of SLAC, was coming to a similar conclusion by a different route. Since the early 1960s, Richter [whose brother, Charles, is a geologist and author of the Richter earthquake scale] had been working on the design of an electron-positron storage ring that could take particles from the linear accelerator and, after storing them, cause them to collide head-on. By 1972, just 21 months after construction funds became available, this facility was in operation. A sketch of the storage ring is shown in Illustration 83, below. Electrons and positrons of energies up to 3 GeV were stored in the ring where, because of the fact that they have opposite electrical charges, they circulated in opposite directions. The magnets around the ring were adjusted to keep the particles moving in their respective paths.

The two circulating beams were allowed to collide in the interaction region, and then counters placed around this region could be used to see what was produced. Schematically, we are dealing with a situation in which an electron and a positron collide and produce an assortment of hadrons. In general, the reaction will look like the one on the left of Illustration 84—a spray of pions, K-mesons, electrons, and positrons will be seen. But, when the experimental evidence started to accumulate, a "bump" was found in the electron-positron cross section, which in-

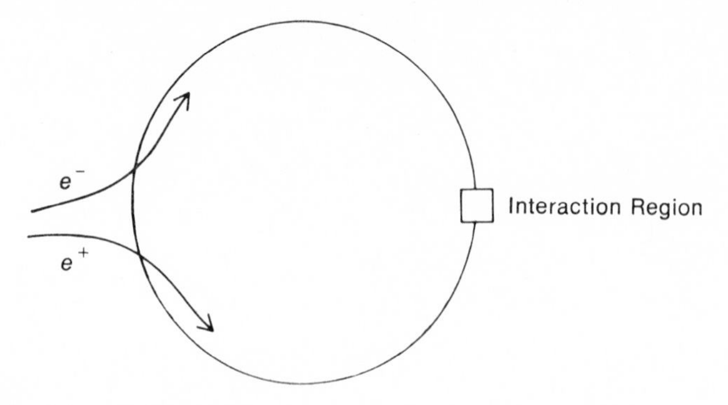

83. Sketch of the storage ring.

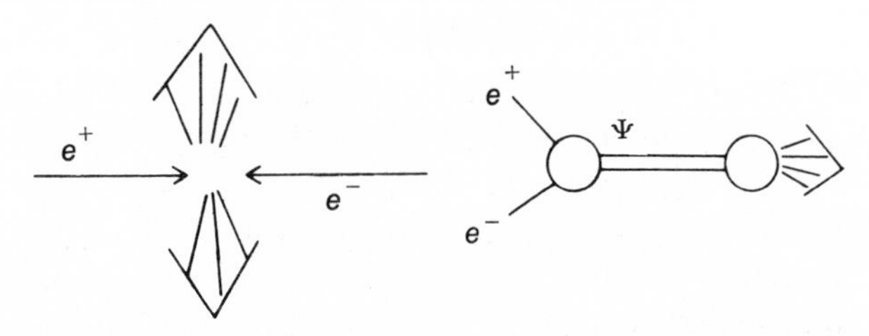

84. *Left:* The production of hadrons. *Right:* The production of the ψ particle.

dicated that the interaction was taking place via an intermediate particle, as seen on the right in the illustration. The Stanford group named their new particle the Ψ.

The two groups learned of each other's work in the fall of 1974, when Sam Ting paid a visit to Stanford. By that time, both groups were ready to announce their results. Two completely independent experiments, using completely different techniques, had discovered the same particle at about the same time. The particle is now called the Ψ/J, and its properties are listed in the table. What these properties mean will be discussed in the next section.

PROPERTIES OF THE Ψ/J	
Mass	3,098 MeV
Spin	1
Isospin	0
Parity	−
Width of resonance	67 keV
Strangeness	0

The Idea of Charm: An Extra Quark

In Chapter XI we saw that one of the major pieces of evidence for the quark model was that by it every known particle was accounted for. This, in turn, implied that there were no "vacancies" in the hadronic scheme of things into which the Ψ/J could be fit. The discovery of a new particle clearly showed that some major overhauling of the quark model was needed—or that it had to be scrapped.

A hint at a resolution of this problem can be seen by considering the width of the Ψ/J. In Chapter VII we saw how the uncertainty principle could be used to relate resonance width to a lifetime for a particle. A width between 100 and 200 MeV, for example, corresponds to a lifetime typical of the strong interactions, such as, 10^{-23} second. The Ψ/J has a width 1,000 times smaller than this, so its lifetime must be over a thousand times greater, or around 10^{-20} second. As far as the strong interactions are concerned, the Ψ/J is virtually a stable particle, like the lambda and the sigma.

This analogy suggests a solution to the puzzle posed by the discovery of the ψ/J. Just as the slow decay of the lambda and the sigma is associated with the presence of the strangeness quantum number and the *s* quark, perhaps the slow decay of the Ψ/J is associated with a new quantum number and a new kind of quark. In that case, the general rule for decays we discussed in Chapter IX would tell us that converting the new quark into a *u* or *d* could not be done in 10^{-23} second, but would have to involve a time similar to that required to make the same conversion for the *s* quark.

As it happened, theorists had suggested that a fourth quark might exist. The argument was based on some concepts of symmetry between hadrons and leptons that seemed to require that quarks come in pairs. The *u* and *d* quarks clearly do this, but the "partner" of the *s* quark had been missing when the prediction was made. In addition, the existence of such a quark would explain certain properties of weak interactions that had been puzzling up to that time. Thus, when the Ψ/J was discovered, a theoretical groundwork had already been laid for an addition to the roster of quarks.

The new quantum number had been labeled *charm,* and was denoted by the letter *C.* The quark that carried it was called the *c* (for charmed) quark. The relation between charm and the *c* quark is exactly analagous to the relation between strangeness and the *s* quark. A particle that contains one *c* quark would have a charm of +1, a particle containing two *c* quarks, a charm of +2, and so on. Each time a *c* quark has to be converted to a *u* or *d* quark, the corresponding particle reaction takes a time long compared to the time associated with the strong interactions.

All of the particles we considered before the discovery of the Ψ/J were formed from *u, d,* and *s* quarks and, hence, had $C = 0$. This is similar to the fact that the nucleon and delta families, being made from *u* and *d* quarks, have $S = 0$. Like the other quarks, the *c* quark has spin $\frac{1}{2}$ and baryon number $\frac{1}{3}$. It has strangeness 0, $C = 1$, isotopic spin 0, and charge $\frac{2}{3}$. Knowing these facts, we can use the techniques of the ordinary quark model to determine what new kinds of objects we should see when we open the charm dimension in the elementary particle world. But before we start this job, let us consider the ψ/J itself.

Since it is made from an electron-positron interaction, it must have $B = 0$, and hence be a meson. It must, therefore, be composed of a quark-antiquark pair. And since all the states involving *u, d,* and *s* quarks are used up, the process of elimination leads us to the conclusion that the ψ/J is made up of a *c* quark and an anti-*c* quark as pictured in Illustration 85.

If this is the case, then two predictions can be made. First, there ought to be other particles with $C = 0$, corresponding to the *c* and $\bar{c}$ in different orientations. Second, there ought to be particles that have $C = 1$, corresponding to a *c* quark and an anti-*u* or *d* quark. Both of these predictions have been borne out by experiment.

The first point was quickly perceived by theoretical physicists. If the ψ/J is really a c–$\bar{c}$ system with the quark spins aligned, there ought to be another particle corresponding to the quark spins antialigned, and still another corresponding to the situations where the *c* and $\bar{c}$ are moving around each other in orbits. In short, the theory predicts that there ought to be many particles, each decaying into lower mass members of the group by emitting gamma rays. It should be possible to detect these decay gamma rays in the laboratory. They were, in fact,

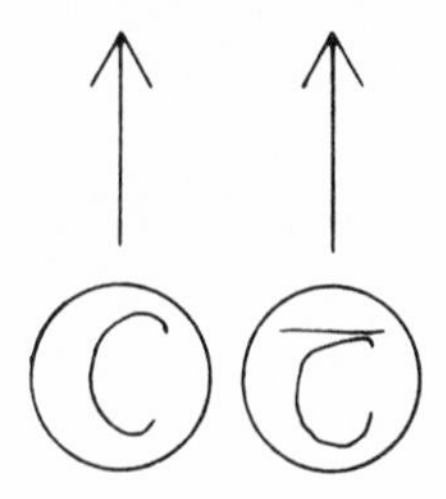

85. Composition of the ψ/J particle.

seen at accelerators in Hamburg and in Italy (as well as at SLAC) shortly after the discovery of the ψ/J itself. In Illustration 86 we show the present c–$\bar{c}$ family of particles as they are known as present. Obviously, the fact that all of these predicted companions of the ψ/J exist lends a lot of credibility to the theory of a fourth quark.

A much more direct bit of evidence would be the existence of a particle that had $C \neq 0$—a particle, in other words, that contained a single c quark. There ought to be mesons, for example, like the one in Illustration 87, in which a c quark is matched with a $\bar{u}$ quark to produce a particle with $Q = 0$ and $C - 1$. The same theoretical considerations that led to the prediction of charm in the first place indicate that this meson ought to decay into a K-meson and a few pions. Thus, it is a straightforward matter to look at the electron-positron reactions that give rise to K-mesons and pions to see if any peaks are seen in the phase space diagram. In 1976, data similar to those in Illustration 88 were obtained from the storage rings at SLAC. They show the number of events leading to a final state of a K- and pi-meson plotted against the energy of the mesons. Clearly, there is evidence for a particle there—a particle that was named the D°-meson. It has a mass of about 1.8 GeV and is, in fact, the particle whose quark structure we pictured above. Since its discovery, other particles in the D family have been found, as would be expected. These include a D* (similar to the D°, except that the quark spins are aligned) and a set of charged D-mesons made from various combinations of c and u quarks.

From these developments, it is clear that the addition of the charmed quark simply adds a new dimension to the quark model, but does not change it in any essential way. The D*, for example, bears the same relation to the D° as the rho-meson does to the pi-meson. All of the other games we learned to play in constructing particles from quarks can be played with the addition of charm. For example, there ought to be particles (made from c and s quarks) that are both charmed and strange. These have not yet been found, but there is every reason to expect that they will soon be seen. We would also expect to see baryons made with one, two, or three c quarks in combination with the others. In fact, there should be a whole new set of particles waiting to be found in accelerator experiments—enough to keep experimental physicists happy for a long time. As an example of how these newly predicted particles might look in a standard eightfold way plot, we show in Illustration 89, on page 182, the predicted baryons associated with the spin

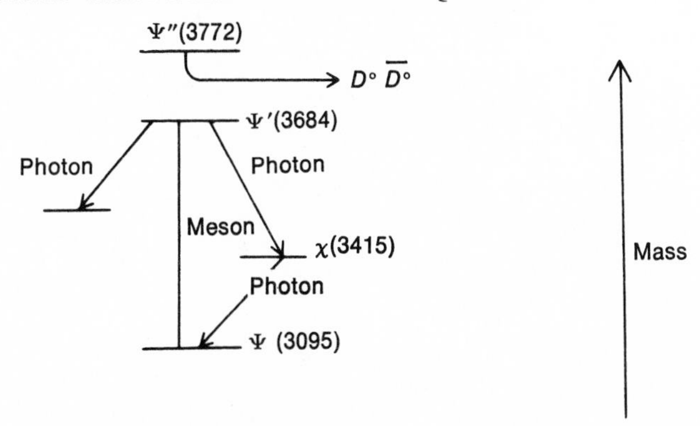

86. The currently known c–$\bar{c}$ family of particles.

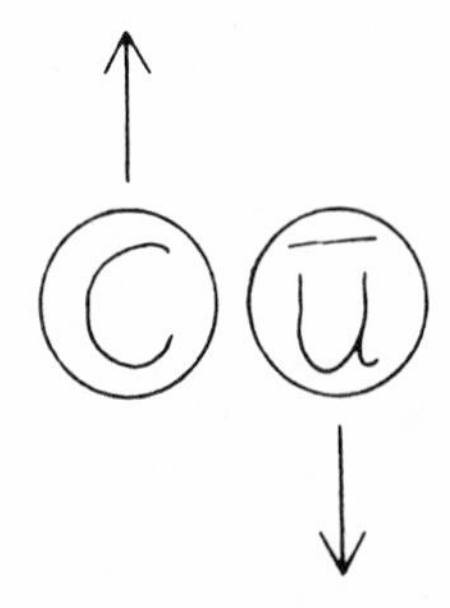

87. The $D°$-meson, which shows charm directly.

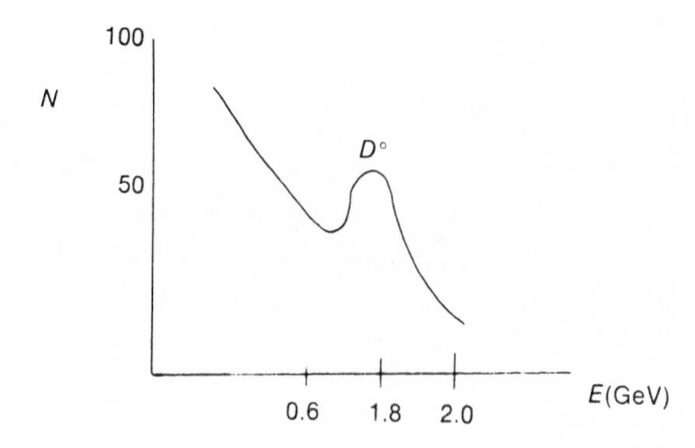

88. The bump in the cross section corresponding to the $D°$.

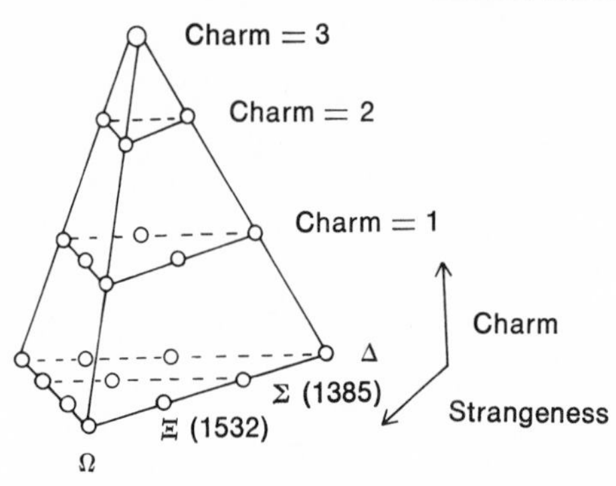

89. Standard eightfold way plot showing predicted baryons associated with the spin 3/2 baryons.

3/2 baryons. This is, you will recall, the grouping of particles that led to the prediction and discovery of the omega—one of the first great triumphs for this approach to elementary particles.

It should be understood that being forced to add charm to our list of quarks suggests a disturbing possibility. What if it is not the last quark to be forced on us, but only one of many? If this were so, then the original hope of being able to use the quark model to create a simple picture of nature would be in serious trouble. Charm by itself, of course, does not imply any such conclusion. A world composed of four basic particles is as simple as a world made up of three. But remembering that the proliferation of chemical elements and elementary particles is what led to the idea of quarks in the first place, we recognize that any hint of a proliferation among the quarks must be taken very seriously indeed.

More Quark Candidates

IN 1977 a group under the direction of Leon Lederman at the Fermi National Laboratory near Chicago announced a result that was to have profound implications for the quark model. The experiment they performed was similar in design to the Brookhaven experiment that led to the discovery of charm. A beam of high-energy protons from the accelerator was allowed to strike a nuclear target (either copper or platinum), and then detectors were set up to look for pairs of oppositely

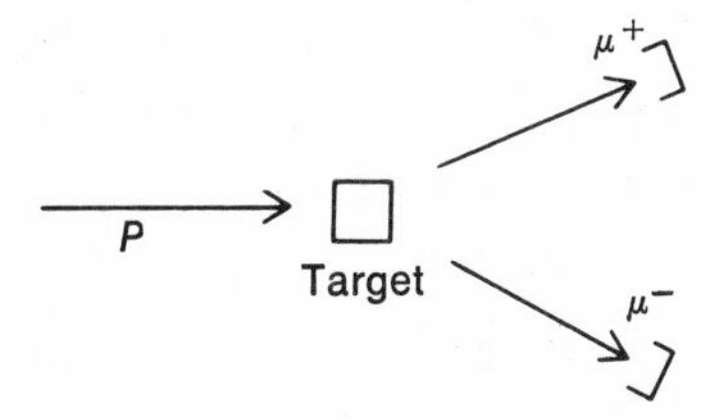

90. Apparatus for experiment to find oppositely charged mu-mesons.

charged mu-mesons. (See Illus. 90.) The main differences between this experiment and the one that Ting had carried out were (1) that the Fermilab accelerator has a much higher energy proton beam, and (2) that the detection of mu-meson pairs rather than electron-positron pairs makes certain parts of the work a little easier to do.

In a sequence of events that seems almost a replay of the 1974 discovery of charm, this experiment produced data for the number of events in which a muon pair is seen as a function of the energy of that pair. (See Illus. 91.) Again, a series of bumps in the otherwise smoothly falling data indicated the presence of a new particle (or set of particles). As before, there was no place for these "extra" particles in the quark model, and, as before, there were theoretical arguments based on studies of weak interactions that suggested that there ought to be a new kind of quark.

Data from Fermilab and other laboratories have pretty well shown that there are, in fact, two particles causing wiggles on the curve in Illustration 91, and there may be others as well. These particles are denoted by the Greek letter Υ (upsilon) and Υ′; they have masses of 9.5 and 10.0 GeV, respectively. These particles are now universally ac-

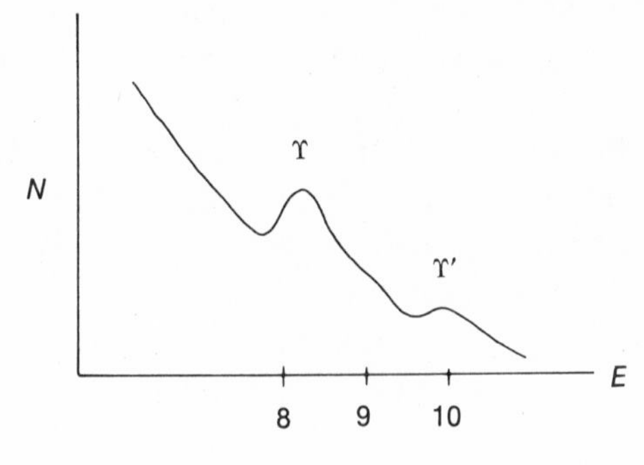

91. Graph showing probable existence of the Υ and Υ′ particles.

cepted as being states similar to the ψ/J, in which a new quark and its antiquark are bound into a meson. The new quark is called the *b,* and the upsilon particle is constructed as follows (Illus. 92):

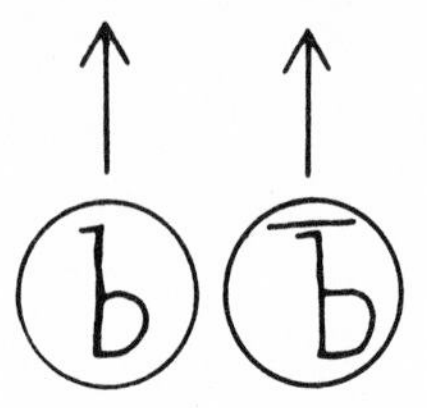

92. Construction of the upsilon particle.

The ϒ′ is presumably the same set of quarks with spins aligned differently, and we can expect a set of new particles to be found corresponding to the ones that were discovered after the ψ/J. Eventually, we can expect to see particles created that exhibit the *b* quantum number explicitly, just as the *D*-mesons were found to exhibit charm. Hence, the discovery of the ϒ seems to start the whole process of building up new particles from a new quark that we went through with charm, and all of the comments we made in the previous section should apply here as well.

The theoretical arguments that suggested the existence of another quark besides *u, d, s,* and *c* actually predict a pair of quarks. These are labeled *b* and *t.* The *b* has charge $-1/3$ and the *t* has charge $+2/3$. Otherwise, each is similar to the *c* quark, except that it carries a new quantum number. Although a valiant effort was made to name them "beauty" and "truth," Keats eventually lost out to the more prosaic *bottom* and *top.* Thus, we would say that the ϒ particle is made of a bottom quark and antibottom quark, and that eventually we will see particles that explicitly exhibit the bottom quantum number. As I write (spring 1979), there is no evidence to indicate that particles made from a top quark have been seen, and there is no guarantee that they can be produced in the current generation of accelerators. However, historical precedent certainly seems to indicate that we shall eventually see such particles. The most likely sequence would seem to be the appearance of a narrow bump in a cross section, followed by a rapid proliferation of states and the eventual discovery of particles carrying new quantum numbers.

We expect that in the near future we shall be confronted with six different kinds of quarks. For the sake of convenience, we list their properties in the following table:

QUARK	SPIN	CHARGE	OTHER QUANTUM NUMBER
u	$\frac{1}{2}$	$\frac{2}{3}$	none
d	$\frac{1}{2}$	$-\frac{1}{3}$	none
s	$\frac{1}{2}$	$-\frac{1}{3}$	$s = -1$
c	$\frac{1}{2}$	$\frac{2}{3}$	$c = +1$
b	$\frac{1}{2}$	$-\frac{1}{3}$	$b = +1$
t	$\frac{1}{2}$	$\frac{2}{3}$	$t = +1$

All, of course, have baryon number 1/3.

Theoretical Proliferation: The Idea of Color

IN the preceding sections we have seen how experimental results have forced us to increase the number of quarks we want to call basic. While this process was going on, a parallel development was taking place in the theory of quarks which, in essence, had the same effect. To understand the reasoning, we can go back for a moment and think about an ordinary atom.

The first three elements of the periodic table—hydrogen, helium, and lithium—are shown in Illustration 93. Hydrogen has a nucleus composed of a single proton, and a single electron moves in the lowest orbit. Helium has a nucleus made up of two protons and two neutrons, and, hence, must have two electrons. Both of these can be found in the lowest orbit. Lithium has a nucleus composed of three protons and four neutrons, and must therefore have three electrons. From the diagram, we see that two of these electrons, like the two electrons in helium, occupy the lowest orbit, but the third electron resides in the next higher orbit. It is almost as if the first two electrons "fill up" the lowest orbit, so that the third has to find room for itself elsewhere. The idea that electrons can fill up orbital spaces follows from something called the Pauli exclusion principles. Although the derivation of this law from quantum mechanics is rather abstract, the rule itself can be stated quite simply: *No two identical spin 1/2 particles can be in the same state.*

One way to perceive this principle is to imagine that each orbit in an

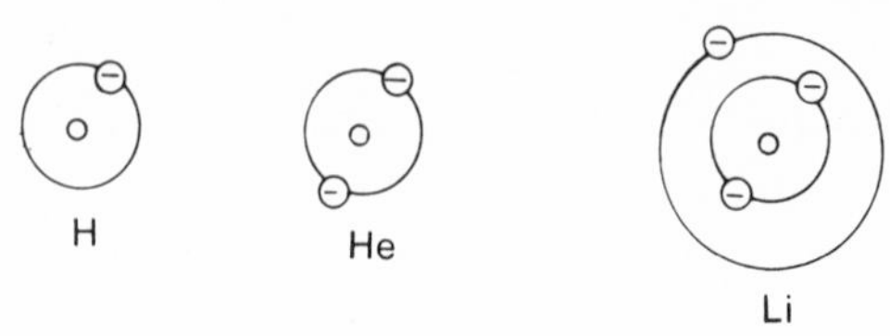

93. Hydrogen, helium, and lithium.

atom is a kind of parking lot for electrons. Each electron in the orbit occupies one space, and when all the spaces are taken, the parking lot is full. Additional electrons must then go to the next highest orbit in which the parking lot is still empty.

Perhaps the easiest way to see how the Pauli principle works in practice is to consider electrons in the lowest orbit. The "state" of the electron in that case corresponds to the direction of its spin. Since this can have two values (up or down), we say that there are two electron states in the lowest orbit or, from the parking lot analogy, that there are two "parking spaces" there. This means that the orbit will be completely filled when it contains two electrons with their spins oriented as shown in Illustration 94. The Pauli principle guarantees that in the lithium atom, the third electron must be in some other orbit.

In passing, we should note that the same laws of quantum mechanics predict that there should be eight states in the next orbit. Since the chemical properties of an atom depend on the outermost electrons, this is the explanation of the periodic table of the elements we talked about in Chapter IX. There are only two elements (helium and hydrogen) in the first row of the periodic table. Lithium, the third element, starts a new row and is placed in the same column as hydrogen. They are chemically similar because they both have a single electron in the outermost orbit. Similar reasoning predicts that the next such element (sodium) should have eleven electrons—two in the lowest orbit, eight in the next, and one in the third. These kinds of electron distributions

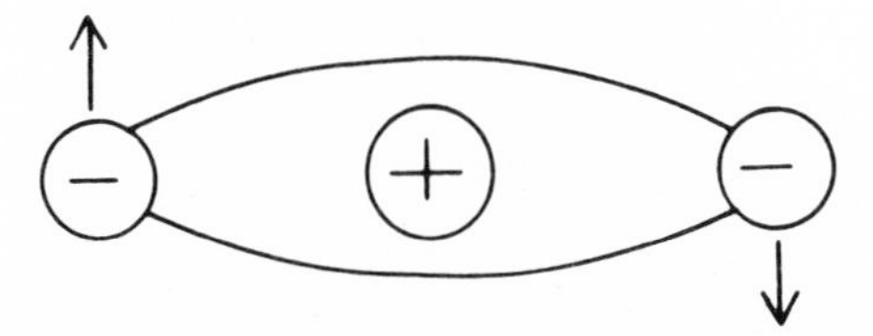

94. Filled electron orbit.

account for the entire structure of the periodic table.

Since quarks are spin 1/2 objects, there is a strong prejudice on the part of theoreticians to assume that they, too, must obey the Pauli principle. Yet there is at least one elementary particle whose very existence seems to imply that they cannot. In Chapter IX we saw that the doubly charged delta (Δ^{++}) was constructed from quarks as shown in Illustration 95—three *d* quarks with spins aligned were required to match the spin 3/2, $Q = 2$ that we observe in this particle.

If the quarks obey the Pauli principle, this particle should not exist, since all three are identical spin 1/2 particles and all three are in the same state. There are only two ways out of this dilemma—either quarks do not obey the Pauli principle, or the quarks in the Δ^{++} are not identical.

If we explore this last possibility, we have to suppose that the reason the Δ^{++} can exist is that its three constituent quarks are not identical in the sense implied by the Pauli principle. Since they have the same charge, spin, parity, and so on, the difference between them must lie in some characteristic that we have not yet discussed. Suppose, for example, that some subatomic gremlin had come around and painted the quarks in three different colors. Then, provided that the three quarks in the Δ^{++} were different colors, there would be no problem with the Pauli principle. Although the three are all in the lowest state

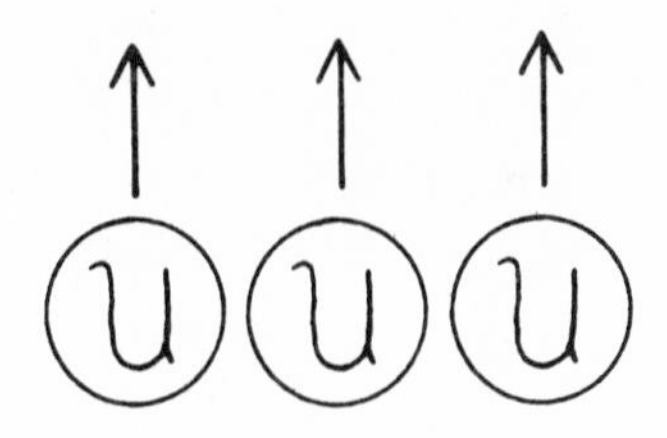

95. Δ^{++}, in terms of quarks.

and spin up, they would not be identical particles because they would have different colors.

The idea of *color* was introduced (under another name) by O. W. Greenberg at the University of Maryland almost as soon as the quark model was put forward in 1964, but it is only fairly recently that the concept has become widely accepted. The delay was inevitable because no particles have ever been seen that exhibit the color quantum number explicitly. Had they been seen, they would have been recognized in the same way that charm and the *b* quantum number were. This amounts to a rule which states that although individual quarks have color, the net color (a quantity derived from the sum of the colors of the quarks) of known particles must be zero.

At this point, we ought to pause for a word of caution. When I introduced the idea of strangeness, I commented on the problems that physicists create for themselves when they use common words to designate new properties of particles. The term "color" is a perfect example. No one thinks that there is any gremlin running around with brushes and buckets, daubing paint on the quarks. No one really thinks that it is necessary to picture quarks as little spinning spheres of different colors. Yet, such a picture is encouraged by the adoption of the term.

For the record then, the term color as it is applied to quarks is not the same thing as the term color as it is applied to everyday objects. In physics it refers to a property of the quarks in the same way that spin, parity, strangeness, and charge signify properties. And just as we can determine the charge of a particle by adding up the charge of each constituent quark, we can find the color of a particle by combining the colors of each quark in it. The rules for combining color turn out to be a little more complicated than the simple addition that we apply to charge, but the principle is the same.

The rule for making hadrons from colored quarks amounts to requiring that when the colors of the quarks are added the result is zero net color. If there were a similar rule for electrical charge, it would require that only those particles for which the charge of the quarks added to zero could actually exist, and if this were true, we would only see electrically neutral particles in nature. No such rule exists for charge, of course, but it seems to exist for color. As a consequence, only particles with zero color are actually seen in the laboratory.

There is some experimental evidence that color actually is a property of quarks. Two experimentally measurable quantities depend on the number of quarks. One of these is the rate at which the π° decays into two photons. The other is the probability that electrons and positrons will create hadrons when they collide. Both these numbers seem to indicate that there are three times as many quarks as would be expected without color. These results would be in agreement with the model if each different color for a quark were counted separately.

In addition, the rule about only zero color (or "color neutral") objects being seen in nature also has a relation to the theory of quark confinement discussed in Chapter XI. If the individual quarks have color, then the zero color rule would say that individual quarks cannot be seen in the laboratory. Therefore, unless the rule can be broken, we should never expect to see quarks except in those combinations that have zero color; that is, in the combinations that give us the particles we already know about.

A Summing Up

THE quark model was originally introduced to bring simplicity to a situation in which the number of elementary particles was expanding beyond desirable bounds. Originally, it seemed to fulfill this function. Reducing the world to three basic building blocks seemed to be a realization of the quest for simplicity. Yet by this time you are probably feeling the impression of déjà vu twice over. Just as the number of chemical elements proliferated, and just as the number of elementary particles proliferated after that, the number of quarks since 1974 has also proliferated. In fact, some physicists refer to the announcement of the discovery of the ψ/J as the "November Revolution," because of what it started.

At the moment there is clear evidence for the existence of five different kinds of quarks. In addition, some theorists predict that a sixth (the *t*) will turn up soon. If each of these comes in three different colors, that makes a total of eighteen quarks. Most of us would agree, I suspect, that even if the quark model is ultimately vindicated, events have already progressed beyond the point where we can speak of explaining the universe in terms of a small number of constituents. Eighteen is hardly a small number in this sense.

Nor can we be sure that the number of quarks will stop with eighteen. There is no reason why more quarks such as the *c*, *b*, and *t* will not turn up in future experiments. If we stop to think, we realize that we already have more quarks in our theory than we had hadrons in the early 1950s, when the study of high-energy physics started. Either the quarks are themselves manifestations of some yet more fundamental entities, or our search for ultimate simplicity is a chimera. Twenty years from now we may know the answer to this question; for the moment, we simply conclude that the quark model has developed some troublesome complexities of its own.

One response to this complexity has been the introduction of another rather whimsical term to describe quarks. We have been speaking of the "kind" of quark, but it is clear that two things are needed to describe a quark. First, we must specify whether we are talking about the *u, d, s, c, b,* or *t* type, and then we must specify its color. The first property—the one that tells us whether we are dealing with a *u* or *d* or *s* or some other—is called the *flavor* of the quark. In modern terminology, we say that quarks now come in six flavors, each of which can have one of three colors.

With this development, we have come about as far as we can on the road to simplicity in detailing the structure of hadrons. We now turn to a short discussion of the parallel development of the ideas about leptons and the weak interactions.

CHAPTER XIII

Leptons and the Weak Interactions

As I was going up the stair
I met a man who wasn't there.
He wasn't there again today.
I wish, I wish he'd stay away.

—HUGHES MEARNS,
"The Psychoed"

The Discovery of the Neutrino

IN Chapter II we discussed the beta decay of the neutron as an example of a weak interaction. We saw that in order to preserve the laws of energy and momentum conservation in this decay it was necessary to assume the presence of an unseen particle in the interaction—a massless, uncharged particle that was called the neutrino. We also saw that this hypothetical particle was so difficult to detect that it could quite literally pass through a block of lead several light-years thick without disturbing a single atom. But so successful was the theory of beta decay which Enrico Fermi put together that physicists were willing to accept the neutrino as a genuine particle despite the fact that it had never been seen in a laboratory. Indeed, it would not be an overstatement to say that there were probably a good number of physicists in the early 1950s who had more faith in the existence of the neutrino than they had in the new strange particles that were starting to turn up in cosmic ray experiments.

In 1956 this faith was justified when two physicists, Clyde L.

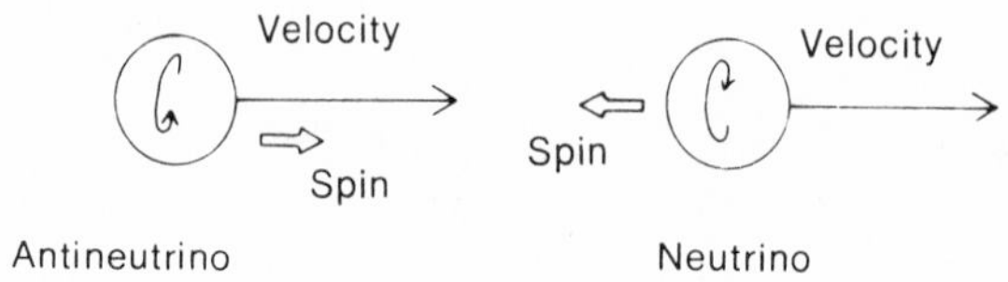

Velocity
Spin
Antineutrino
Velocity
Spin
Neutrino

96. Spin directions of the antineutrino and neutrino relative to their velocities.

Cowan, Jr., and Frederick Reines of Los Alamos, managed to produce laboratory evidence showing that the neutrino also existed in the real world, and not solely in the minds of theoretical physicists. Considering the miniscule effects that the neutrino has on its surroundings, this was no mean feat. It took five years of searching and refining the experiment before Cowan and Reines finally announced a definitive result.

Before going into the details of the experiment, we must digress briefly in order to define some terms. We have seen that an elementary particle can be characterized by its spin, and that the direction of the spin obeys the so-called right hand rule (see Chapter VIII). We have also seen that for every particle in nature there is an antiparticle. Consequently, we can conclude that there ought to be an antineutrino, which differs from the neutrino in the direction of its spin (since the neutrino has no charge, spin is the only property that can be different in the particle and antiparticle). By convention, the particle whose spin is in the same direction as its velocity is called the antineutrino, while the particle whose spin is directed opposite to its velocity is the neutrino. This is illustrated in Illustration 96, above.

If we use this convention, then the particle that is the invisible partner in the beta decay of the neutron is actually the antineutrino. It is customarily denoted by $\overline{\nu_e}$, where the subscript e refers to the fact that it is produced in concert with an electron. In keeping with this convention, we now write the neutron beta decay as $n \rightarrow p + e^- + \overline{\nu_e}$.

Since the difficulty of detecting the presence of a neutrino or antineutrino arises because of the very small probability that the particle will interact with the nuclei it passes, the only way to see such interactions is to find a source of the particles copious enough to overcome the small probability of interaction. You can see how this line of reasoning works by noting that the expected number of interactions that will be seen each second will be

events seen = number of neutrinos or antineutrinos ×
probability of an interaction

The fact that the probability of interaction is small can be overcome if the number of neutrinos or antineutrinos is large enough.

Nuclear reactors, as a by-product of fission, produce large numbers of antineutrinos—perhaps as many as 10^{18}/second. By placing a large target and counter apparatus near the reactor at Savannah River, South Carolina, Cowan and Reines calculated that they should be able to get one reaction of the type

$$\overline{\nu_e} + p \rightarrow n + e^+$$

every 20 minutes. Not a prodigious number of events, of course, but enough to do the job.

The apparatus they used is sketched in Illustration 97. It consisted of layers of water (which contains hydrogen atoms whose nuclei served as the target for the antineutrino projectiles) interleaved with layers of a scintillating liquid. The entire block was then surrounded by detectors that would see the photons given off in the scintillators.

When one of the rare antineutrino interactions occurs in the water, there are two products—the positron and the neutron. The positron will annihilate with an atomic electron in the water within a millionth of a second or so. When this happens, two energetic photons are emitted, each of which enters a scintillating layer and produces a shower of photons that will be seen by the detectors. The neutron does not interact so quickly, since it has no electrical charge. Instead of losing its

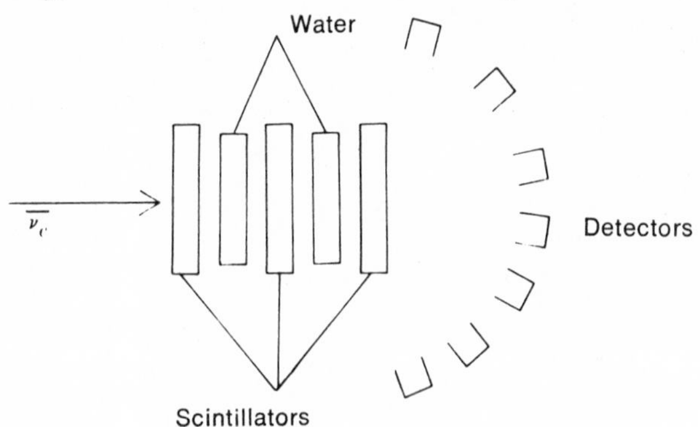

97. Apparatus used to find the antineutrino.

energy through an electrical force, it undergoes a series of collisions that slow it down to the point where it can be captured by a nucleus. The water in the target is mixed with a small amount of cadmium chloride to augment the absorption process, cadmium being an element that absorbs slow neutrons very efficiently. When the neutron is absorbed in the cadmium, the new nucleus gives off one or more photons as the protons and neutrons rearrange themselves to accommodate their new partners. These photons are also converted into a signal in the scintillator.

Thus, the sequence of events that would signal the presence of an antineutrino interaction would be (1) two photons from the electron-positron annihilation, followed in a few millionths of a second by (2) one or more photon characteristic of the cadmium nucleus. When Cowan and Reines finally satisfied themselves that they were seeing these events at the expected rate of one every 20 minutes, the physics community had the proof it needed for the existence of the antineutrino—a proof the physics community had waited for for 20 years.

The Mu-Neutrino and a New Conservation Law

ONCE the antineutrino associated with beta decay had been seen in the laboratory, it was only natural that attention should turn to the problem of providing neutrino beams in accelerators. As we discussed in Chapter VI, this can be done as shown in Illustration 98. A beam of pi-mesons of known charge and energy is taken from an accelerator and allowed to decay into a muon and a neutrino. Our convention on neutrinos and antineutrinos tells us that the two decay reactions are

$$\pi^+ \rightarrow \mu^+ + \nu_\mu$$

and

$$\pi^- \rightarrow \mu^- + \bar{\nu}_\mu$$

where the subscript μ is used to remind us that these particular neutrinos are associated with the production of the mu-meson. If the mixed beam of muons and neutrinos is then taken through a large shield (e.g., a pile of steel plates or a long earth mound), we can, by choosing a thick enough shield, ensure that the muons and any other particles that

98. Production of a neutrino beam in an accelerator.

might be in the beam by accident will be absorbed in the shield and only the neutrinos will pass through unaffected.

In 1962 a group at Columbia University reported on experiments designed in that way. Their purpose was to look for reactions of the type

$$\bar{\nu}_\mu + p \rightarrow n + \mu^+$$

and

$$\nu_\mu + n \rightarrow p + \mu^-$$

which would be initiated by the neutrinos and antineutrinos. The signature of such an event would be the sudden appearance in the apparatus of a charged particle (the muon). Both the incident neutrino and the final neutron, being uncharged, would be invisible.

The most striking feature of the experiments was that *only* the reactions given above were observed. No reactions of the type

$$\nu_\mu + n \rightarrow p + e^-$$

and

$$\nu_\mu + p \rightarrow n + e^+$$

were seen at all. In other words, it appears that neutrinos resulting from the decay of muons cannot create electrons or positrons, even though the Cowan-Reines experiment showed conclusively that neutrinos emitted from beta decay could do so. There are, in other words, two different kinds of neutrinos, one associated with electrons and the other with muons.

This, of course, is the reason we referred to neutrinos as ν_μ and ν_e. Not only does each neutrino appear in association with a particular lepton, but it can initiate reactions only if that same lepton is involved. It seems, therefore, that the leptons come in pairs, the electron with its neutrino and the muon with its neutrino.

The two neutrinos also cast some interesting light on a conservation law. By our choice of definition of the neutrino and antineutrino, we have arranged things so that there is a law for weak interactions similar to the conservation of baryon number. This is the law of conservation of leptons: *The net number of leptons cannot change in any interaction.*

For example, in neutron beta decay we have no leptons in the initial state, and an electron with its antineutrino in the final state. If we assign a *lepton number* of +1 to the electron, then we have to assign −1 to the antineutrino (just as we would assign a baryon number −1 to an antiproton). The net lepton number after the decay is therefore zero, the same as it was initially. You can, if you check the other weak interactions we have discussed so far, see that this law holds.

The existence of the two neutrinos suggests further that a more stringent form of this law may apply. If we assign an electron number to the electron and its neutrino and a muon number to the muon and its neutrino, then it appears that electron number and muon number are each conserved individually. In other words, if we consider the reaction

$$\mu^- \rightarrow e^- + \overline{\nu_e} + \nu_{\overline{\nu}}\,\mu$$

we start with a muon number of 1 and electron number 0. The neutrinos in the final state are arranged to give the same value for each of these. We then note that if electron and muon numbers are each conserved, the law of conservation of lepton number is automatically satisfied. These concepts are summarized in the table below, in which the four leptons are given with their various quantum numbers.

Two important inferences follow from the two neutrino theory. First, if it turns out that the electron and muon are not the only massive

PARTICLE	LEPTON NUMBER	ELECTRON NUMBER	MUON NUMBER
e^-	1	1	0
e^+	−1	−1	0
μ^-	1	0	1
μ^+	−1	0	−1
ν_e	1	1	0
$\overline{\nu_e}$	−1	−1	0
ν_μ	1	0	1
$\overline{\nu_\mu}$	−1	0	−1

leptons in nature (and we shall see later that they are not), then we would expect a new kind of neutrino to accompany each new massive lepton that is found. Second, if, as some theorists believe, there is some fundamental connection between leptons and quarks, then the fact that leptons always come in pairs (particle plus neutrino) would imply that quarks should always come in pairs as well. This type of reasoning is what led to the first predictions that there ought to be a charmed quark as a complement to the strange quark. Pushing the analogy a bit further led to the prediction of the b and t quarks after another lepton, the τ, was found. This discovery will be discussed later in this chapter.

Parity in Weak Interactions

IN Chapter VIII we introduced the idea that we could describe particles by a property called parity. In essence, the parity of a particle tells us how its wave function will look if the space coordinates are inverted. A simple way to visualize this is to say that the parity operation corresponds to looking at the particle in a mirror. If the reflection is the same as the particle, we say that the particle has positive parity.

Imagine this mirror-image operation being applied to processes as well as to particles. For example, a collision in which a proton approaches from the right and an electron from the left, when looked at in a mirror, would appear to be a collision in which a proton approached from the left and an electron approached from the right. Intuition tells us that what happens in the collision should not depend on whether we view the process directly or see it in a mirror. In physics, the idea that this particular operation should not affect anything in nature is called the principle of parity invariance. The principle holds for strong and electromagnetic interactions, and, until the early 1950s, was believed to hold for weak interactions as well.

In 1956 two young physicists at Columbia University, Tsung-Dao Lee and Chen Ning Yang, were studying some problems in weak interactions. As part of their study, they took a close look at the reasons why parity invariance was believed to hold in the weak interactions. After some careful thought, they came to the conclusion that there was no actual evidence that indicated that it must, but that there was simply a strong prejudice on the part of theoretical physicists that the weak interactions had to be like the rest of physics. On the strength of this realization, they modified the theory of beta decay to see what would happen if the idea of parity invariance did not hold. They found that there were a few situations in nature where this lack of mirror symmetry could actually be seen and measured.

In 1956 Madame Chien-Shiung Wu, also at Columbia, performed an experiment that verified beyond question that the weak interaction was not invariant under parity. Although technically difficult, the idea of the experiment was quite simple. A piece of material containing cobalt (actually, the isotope ^{60}Co) was cooled to within a few tenths of a degree of absolute zero ($-273°$ C). At this very low temperature a magnetic field can be applied that will lock all of the cobalt nuclei into an orientation where their spins are all pointing in the same direction. The low temperature is required so that the normal oscillations that atoms perform, and that we perceive as heat, are reduced to the point where they cannot upset the nuclear alignment. Although this procedure sounds simple, it took over 6 months of work to turn the idea of an aligned sample into reality. Even then, the sample could be kept aligned for only 15 minutes, so it could be said that this experiment involved 6 months of preparation for a 15-minute run.

^{60}Co is a nucleus that undergoes beta decay spontaneously, and has a half-life of about 53 years. As is well known, the element is widely used as a source for radiation in cancer therapy in modern hospitals. Madame Wu, however, used the decay property for a different purpose. Since she knew the direction in which the nuclear spin was pointing, she could observe the number of electrons that came off in the "up" direction and the number that came off in the "down" direction. We can understand the significance of this measurement if we look at Illustration 99. At the left is a cobalt nucleus in which the electron is emitted along the direction of the spin—the direction we are calling up; on the right, we show the mirror image of the same process. The spin of the nucleus is reversed, but the electron is still moving toward the top of the page, albeit in a different direction. Thus, if parity invariance is valid in weak interactions, we would expect to see as many electrons emitted along the direction of the nuclear spin (up) as along the opposite direction (down). In 1956 most physicists would have predicted this result.

When the experiment was actually performed the results clearly

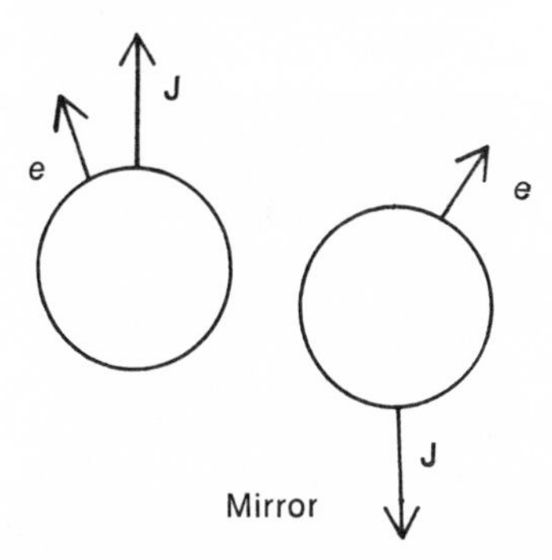

99. Representation of the experiment that disproved the validity of parity invariance in weak interactions.

showed that the electrons were emitted preferentially in the up direction. Thus, as far as the weak interactions are concerned, nature is *not* ambidextrous. There is a differentiation between right and left in a way that is totally unexpected. For predicting that this might be so, Lee and Yang shared the Nobel Prize in 1957—the youngest men every to do so. At the time they actually did the work, Lee was 29 and Yang was 33.

The Hierarchy of Conservation Laws

THE discovery that parity is not conserved in weak processes brings out an important point about the interactions we have studied so far. These have, of course, different strengths (we will come back to this point shortly), but they also seem to be arranged in a hierarchy; that is, the stronger the force, the more conserved properties there are.

For example, the strong interaction does not depend on the electrical charge of the particles being examined. In more technical language, we say that the strong force conserves the isotopic spin in any process. But if we go down one step in interaction strength to the electromagnetic force, this statement is no longer true. The electromagnetic force on a proton is much different from that on a neutron, since the former is electrically charged and the latter is not. Thus, we say that the electromagnetic force does not conserve isotopic spin, or, equivalently, that the electromagnetic force breaks isotopic symmetry.

In the same way, both the strong and the electromagnetic interactions conserve parity, but the weak interaction does not. These differences between the interactions are highly significant, since they tell us something about the way the fundamental interactions operate. In the present state of our knowledge, however, it is not possible to do more than state what the differences between the interactions are, for we have no way of showing why they should be so.

To summarize the properties that are conserved in the three interactions we have discussed so far, we list in the table below a number of conserved quantities and indicate on the right which interaction conserves them.

QUANTITY	STRONG	ELECTRO-MAGNETIC	WEAK
Energy, momentum, angular momentum	↓	↓	↓
Electron number, muon number total electrical charge	↓	↓	↓
Lepton number, baryon number	↓	↓	↓
Time reversal	↓	↓	
Charge conjugation	↓	↓	
Parity	↓	↓	
Strangeness	↓	↓	
Charm	↓	↓	
b and t quantum numbers	↓	↓	
Isotopic spin	↓		

Why Is the Weak Interaction Weak? The W Boson

In Chapter III we saw that the strong interaction could be thought of as being generated by the exchange of virtual pi-mesons. This way of thinking about the interaction is useful, and it led Yukawa to the prediction of the existence of the mesons before they were actually seen in the laboratory. Similarly, we can think of the electromagnetic force as being generated by the exchange of photons. For each of these two forces, there is a particle that can be identified with the force, and which we could say "mediates the interaction."

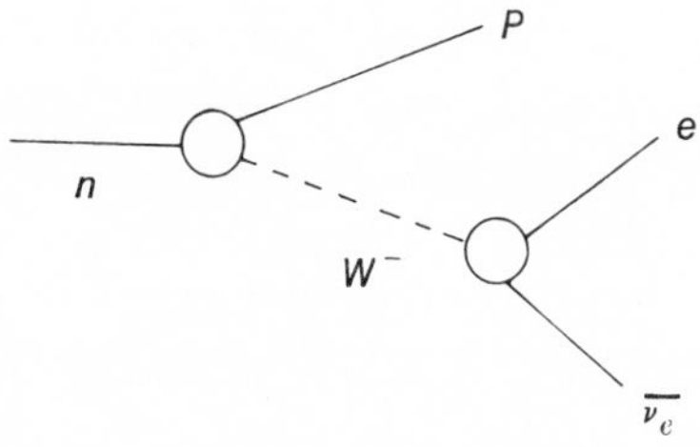

100. Theoretical role of the vector boson in beta decay.

Starting with Yukawa, physicists have often pondered the question of whether such a statement ought not to be true for the weak interaction as well. None of the particles we have discussed so far have the requisite properties to play the role of mediator for the weak interactions, but perhaps there is some other, as yet undiscovered particle, that can. Although this hypothetical particle has never been seen, either directly or indirectly, in the laboratory, it has been given a name. It is called the *vector boson,* or the *W* particle. From the properties of the weak interactions themselves we can deduce a good deal about the properties such a particle ought to have, supposing that it exists at all.

To picture the role that such a particle would play in a process like beta decay, look at Illustration 100. A neutron emits a virtual negatively charged *W* particle, changing into a proton in the process. The *W* then turns into an electron and an antineutrino. This picture, in which the weak interaction is mediated by a virtual particle, puts beta decay on the same conceptual level as strong and electromagnetic processes.

From the diagram, we see that the baryon and lepton numbers of this particle must be zero. Somewhat more technical considerations show that the spin of the particle must be 1. Particles with even spin are sometimes called bosons (after S. N. Bose, who, together with Albert Einstein, first investigated some of their theoretical properties). Particles with spin 1 are called vector particles, so the name vector boson, or "intermediate vector boson," is sometimes used to describe the *W*.

In the "classical" (i.e., pre-1967) theory of weak interactions, the *W* particle had to have either a positive or a negative charge. This requirement was imposed by the fact that in experiments it always appeared that particles undergoing weak decay also underwent a change of electrical charge. In beta decay, for example, the neutron changes into a proton, and the *W* particle that is exchanged has a negative charge.

In 1967, Steven Weinberg of the Massachusetts Institute of Technology published the first paper on a new departure in physics, the idea of a so-called *gauge theory.* It is, as we shall see, one of the most exciting developments in particle physics in a long time. In this new theory, there is a third type of vector boson, one that is not charged. The new particle is called the *Z.* (The experimental consequences of such a particle will soon be discussed; we mention it here for completeness.) In the jargon of high-energy physics, reactions (such as beta decay) in which a charged *W* is exchanged are said to be mediated by a *charged current.* The term current is used here because of the similarity between the two types of processes pictured in Illustration 101. On the left, we show the electrical repulsion between two electrons as it is visualized in particle physics. We say that a photon is exchanged to produce the force, and that the interaction proceeds by the exchange of an *electromagnetic current.* On the right, the beta decay of the neutron is shown in the same way. It proceeds by the exchange of a charged *W* boson, which is called the charged current. A similar process in which a *Z* is exchanged would be said to proceed by the exchange of a *neutral current.*

Perhaps the most promising aspect of the vector boson hypothesis is that it suggests a possible connection between weak and electromagnetic interactions. The idea that there should be distinct types of interactions with different sets of rules governing them is not attractive to physicists. They would prefer to imagine that at some deeper level all these interactions are the same, and that the difference in the way

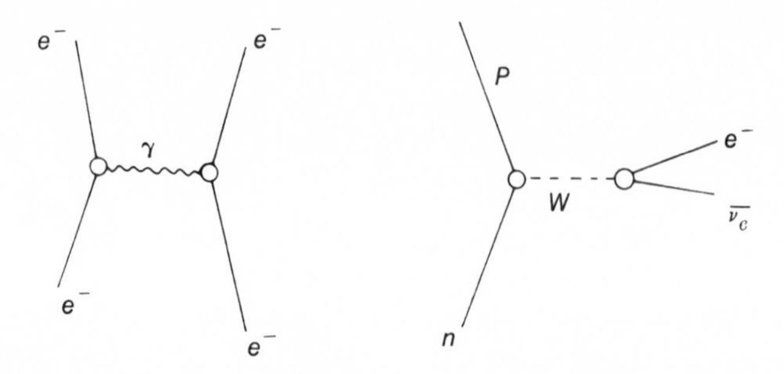

101. Representation of two interactions.

they appear arises from the details of the interaction as it proceeds, rather than from its fundamental nature. We have already seen this kind of reasoning in the concept of isotopic spin. The idea that the proton and the neutron are really the same particle and that they differ only in the orientation of the isotopic spin resulted in an important simplification.

This same kind of simplification may be at hand, thanks to the *W* boson, if the electromagnetic and weak interactions begin to look quite similar. On the left of the illustration we show a typical electromagnetic interaction proceeding by the exchange of a photon. On the right is shown a weak interaction proceeding by the exchange of a *W* boson. The graphs are similar, but we know that in the laboratory the strength of the two interactions differs greatly. One way of explaining this phenomenon is to say that the interactions *are* different. Another is to suggest that the interactions are the same and seem to differ only because they are proceeding through the mediation of different mass particles.

It turns out that the strength of an interaction that is mediated by a heavy particle such as the *W* is the "actual" strength of the interaction divided by the square root of the mass of the particle itself. (The term strength will be given a more precise meaning in the next chapter). Since the weak interaction strength is about 1,000 times smaller than the electromagnetic interaction strength, it follows that the actual strengths of the two interactions will be equal if the mass of the *W* is about $M_W \sim \sqrt{1{,}000} \approx 33$ GeV. This is a very large mass—barely within the range of our current generation of accelerators. But because the existence of the *W* has such important implications for theory, it, like the quark, is one of the entities that is always sought whenever a new accelerator comes on-line.

The Neutral Current and the Heavy Lepton: Two Important Recent Discoveries

THE past 5 years have seen some important advances in the experimental aspects of weak interaction physics. One of these is the availability of a high-energy, high-quality neutrino beam at places such as the Fermilab. The way a neutrino beam is constructed was described earlier. Here I report on how such a beam was used to prove the existence of neutral currents in the weak interactions. If, as in the original theory of weak interactions, there were only charged currents (i.e., only charged *W* bosons), then every interaction initiated by a neutrino would have to look like the one pictured in Illustration 102. A neutrino would come in and emit a virtual charged *W*, thereby changing into a muon or electron, depending on the type of neutrino. The *W* would then interact with the target and produce some final configuration of particles. In this situation, if we looked at neutrino-induced interactions, we would *always* see a charged lepton—either a muon or electron—among the final particles produced. This follows from the conservation of charge at the ν–W–lepton vertex.

If, however, there should exist a neutral current, we could get interactions something like that shown on the left in Illustration 103. A neutrino could emit a virtual *Z* and not change its identity. If we were detecting this kind of event in a device that depended on ionization (such as a bubble chamber), we could not see the neutrino, so the event would look something like the one on the right in Illustration 103. A spray of final particles would suddenly appear in the detector, and there would be no leptons among them.

In 1973 a group working with the neutrino beam at Fermilab announced that they had seen events of this type. Shortly thereafter, similar events were seen in a lower energy beam at the European Center for Nuclear Research (CERN) in Geneva. The neutral currents were there, and this provided an important piece of evidence for the new theories. It also provided an important missing piece to our knowledge of weak interactions.

Another important discovery in lepton physics was made at Stanford in 1975. Since the discovery of the muon in the 1930s and the two neutrino experiment in 1962, it had been assumed that there were four

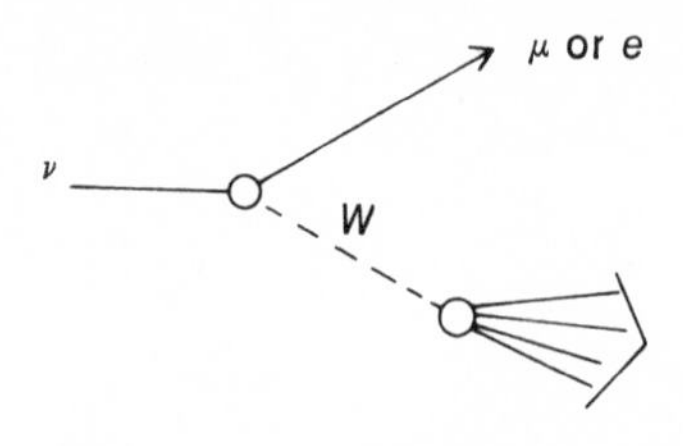

102. The interaction of a neutrino according to the conventional theory.

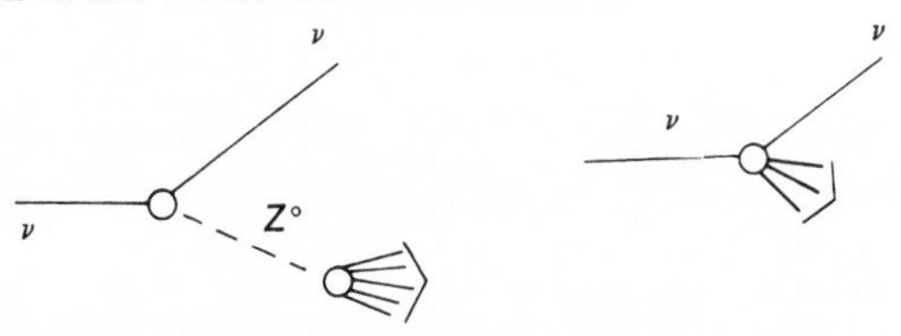

103. Scattering of a neutrino with neutral currents.

leptons—the electron, the muon, and the two neutrinos. This fact was the basis, as we saw, for the theoretical prediction of charm. But the question of whether there were any other leptons more massive than the muon was still left unanswered. The Stanford experiment was a search for such particles. It was performed on the same electron-position storage ring on which the ψ/J had been discovered earlier. What was seen were some 64 events of the type $e^+ + e^- \rightarrow \mu + e +$ undetected particles. According to conventional physics, there is no way such an event could occur. If we imagine that there is a third lepton (we will call it the τ (tau) lepton), then it ought to be able to decay into either the muon or the electron via the interactions

$$\tau^- \rightarrow e^- + \overline{\nu_e} + \nu_\tau$$

and

$$\tau^- \rightarrow \mu^- + \overline{\nu_\mu} + \nu_\tau$$

The ν_τ is the neutrino that (presumably) is associated with the new particle. We could then imagine a reaction like the one pictured in Illustration 104, in which the electron and positron annihilate to form a pair of the new leptons, one of which decays into a muon and the other into an electron.

The report of such events (which, incidentally, was published in a paper with thirty-six authors) is now accepted as evidence for a third lepton. The new particle is called the τ^- and has a mass of 1.8 GeV. In all other respects, it is similar to the muon and electron.

At the present time, therefore, we know of three massive leptons, and no one doubts that a third neutrino associated with the τ^- will be eventually seen. Like the quarks, the number of leptons seems to be proliferating. Just as the idea that there should be a connection between quarks and leptons led to the prediction of charm, the fact that there

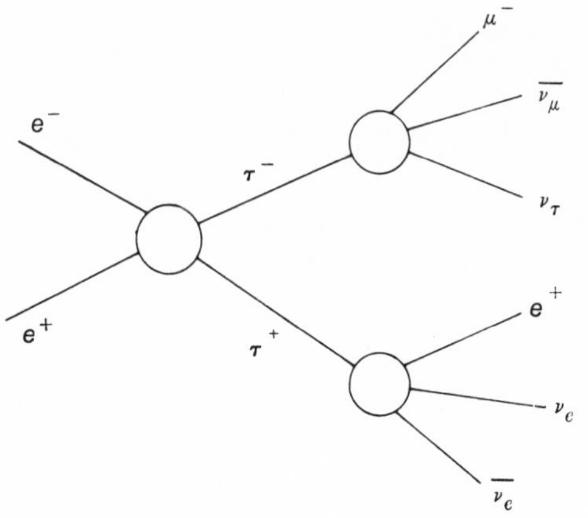

104. Reaction producing the tau-lepton.

are now six leptons has led theorists to suggest that there should be another pair of quarks to match them—the *b* and the *t*. We can therefore expect the next generation of accelorators to detect still heavier leptons, and for each such lepton theorists predict another pair of quarks. This, in turn, will lead to searches for narrow resonances like the ψ/J and the Υ as evidence for the new quarks. The upshot is that the growing complexity of the hadrons has its counterpart in a growing complexity of the weak interactions.

CHAPTER XIV

New Trends and Old Problems

But he grew old—
This knight so bold—
And o'er his heart a shadow
Fell as he found
No spot of ground
That looked like Eldorado.

—EDGAR ALLAN POE,
"Eldorado"

One Dream Dies, Another Is Born: Unified Field Theories

THE prospect that the proliferation of quarks shows the reality of nature to lie at a deeper level than any we have plumbed so far is disheartening. For one thing, in the past it has always required a thousandfold increase in energy to create a new level of particles. We needed energies in the electron-volt range to probe the atom, in the MeV range to probe the nucleus, and in the GeV range to probe the elementary particles. If this is any indication, we shall need energies in the TeV range to probe the quarks themselves, and energies in the thousands of TeV to go beyond that. Given the tremendous cost of building even a 1-TeV machine, it is unlikely that machines of much higher energy will soon be built.

In addition, the stubborn refusal of the quarks to show themselves, coupled with the idea of quark confinement, suggests that even if these energies were available we might never be able to study quarks in isolation. If we cannot, we would be forced to study the putative quark structure indirectly by looking at the elementary particles, and this would be an extremely difficult task.

At the moment, the dream of explaining the entire physical world in terms of a few basic building blocks does not look realizable. Does this mean that we have to surrender the idea that nature is, at some level, basically simple?

Not necessarily. Though things appear to be stalled along the quark front, important new advances are being made in another area—the one dealing with unified field theories. In that domain the search for simplicity is very different from that which we have been following in this book. The road that led us to quarks started with the assumption that nature was simple in the sense that it could be understood in terms of a few simple constituents. We could call this a search for structural simplicity. We can imagine another kind of simplicity, however. We can imagine a world in which we understand processes in terms of a few general principles. The simplicity here would be of an abstract and purely intellectual nature, something like plane geometry, where all the properties of figures follow from a a few basic postulates.

In practice, this search for intellectual simplicity has been concerned not so much with the structure of particles as with their interactions. Thus, instead of concentrating on the search for quarks, the theorist would concern himself with studying the fundamental interactions between elementary particles.

We know, in fact, of only four such interactions. I have mentioned the strong, the electromagnetic, and the weak forces. If we add to these the familiar force of gravity, the list is complete. No matter what happens in the world, the process must involve one or more of the interactions on this list.* The gravitational force is explained by Einstein's general theory of relativity. With these laws we can calculate the gravitational attraction between any two objects, but the theory has nothing to say about the strong, weak, or electromagnetic forces that may exist as well. To study these other forces, we have to go to other theories which, in turn, will tell us how to calculate the electrical forces involved in a given situation, but will be unable to include gravitational effects.

---

*Some physicists have postulated a fifth interaction, which they call "superweak," but so far the evidence does not support this idea.

This type of compartmentalization of the basic forces is troubling to physicists. Confronted with such a situation, theoretical physicists tend to believe that there must be some underlying unity among the interactions, and that the observed differences must be due to some less fundamental details. The introduction of the vector boson (see Chapter XIII) was one result of such a belief. There the assumption was that the weak and electromagnetic interactions are really of the same strength and that they differ only because the particles exchanged in the two interactions, the photon and the *W*, are not the same.

Many great physicists have given much time and thought to framing a so-called unified field theory—a theory that would encompass all four kinds of interactions in a single set of principles. Einstein, for example, spent most of the latter half of his life in an unsuccessful attempt to bring about this sort of unification in physics, as did Werner Heisenberg. The fact that neither of these men was successful suggests the tremendous difficulties involved in formulating a unified field theory.

As I write, however, the vision of a unified field theory is once again alive among theoretical physicists. Before sketching the work being done, it may be helpful to summarize what we know about the four basic interactions.

Each of these interactions can be thought of as occurring through a process something like that pictured in Illustration 105. A force is generated when an intermediate particle is exchanged between two objects. We can think of such an interaction as being governed by two factors: The probability that the virtual exchanged particle will be emitted and the identity of the particle being exchanged. These two properties can be used to determine the observed strength of the interaction.

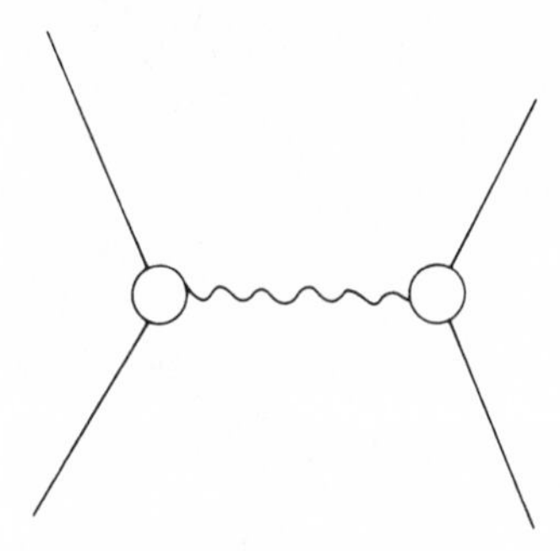

105. Representation of the theoretical process of the four basic interactions.

The probability that the virtual particle will be emitted is called the *bare strength* of the interaction. The actual strength as measured in the laboratory will depend on the bare strength and on the kind of virtual particles involved. Although the connection is somewhat complicated, we can get an idea of how it might arise by recalling the arguments about the range of virtual particles given in Chapter III. We saw that the distance such virtual particles could travel (and, hence, the distance over which a particular force could be felt) was given by

$$d \approx \frac{h}{MC}$$

where M is the mass of the virtual particle. The logic behind the introduction of the W in Chapter XIII was that even though the bare strength of the electromagnetic and weak interactions might be the same, the range of the weak interactions would be very small because of the large mass of the W. Consequently, the actual strength of the weak interactions as measured, for example, by the time scale involved in weak decays, would be small. The hope of the unified field theory is that ultimately, similar connections will be found for all of the interactions.

We are now clear about the exchanged particle for the strong, electromagnetic, and weak interactions. The gravitational interaction is also supposed to be mediated by the exchange of a particle called the graviton. This is a massless particle like the photon, but it has a spin of two, as opposed to spin one for the photon. It has not been detected in the laboratory, but there is little question that it exists.

In the following table we present a summary of our knowledge of the four interactions. The observed strengths are presented on a scale in which the strong interaction has a strength of one.

INTERACTION	OBSERVED STRENGTH	RANGE	PARTICLE EXCHANGED
Strong	1	10^{-13} cm	hadrons
Electromagnetic	$\frac{1}{137}$	infinite	photons
Weak	10^{-5}	10^{-15} cm	W
Gravitation	6×10^{-39}	infinite	graviton

Gauge Theories

IN 1967 Steven Weinberg of the Massachusetts Institute of Technology, followed independently a few months later by Abdus Salam of the International Center for Physics in Trieste, made an important advance toward establishing the unified field theory. Using a mathematical technique called *gauge symmetry,* they were able to show that a theory could be constructed in which the weak and electromagnetic interactions were unified. Thus, the electromagnetic interactions (which proceed by the exchange of a photon) and the weak interactions (which proceed by exchange of the *W* and *Z* intermediate bosons) are both thought of as arising from the exchange of a single family of particles. In this sense, the interactions are the same.

Perhaps a somewhat more familiar example will clarify this idea. We have seen that the nuclear binding force can be thought of as arising from the exchange of pi-mesons. In the nucleus, the neutral pi-meson, the positive pi-meson, and the negative pi-meson all participate in this exchange process. We do not consider the strong force associated with the exchange of the various pi-mesons to be different in any fundamental sense, however, because the mesons differ from each other only in the orientation of their isotopic spins. The forces associated with the neutral pion are thus considered to be identical to the forces associated with a charged pion (except for some minor details). In a similar way, gauge theories view the forces associated with the exchange of a photon (the electromagnetic forces) to be identical with those associated with the *W* and *Z* (the weak forces).

We can get some idea of how a gauge theory works by considering a rather mundane example. Suppose we had a barrel that we knew contained equal numbers of apples and oranges. We could imagine going through the barrel, replacing every apple by an orange and vice versa. After completing this operation, we could ask whether any properties of the barrel had changed. More particularly, we could ask what restrictions would be placed on the individual apples and oranges by the requirement that a given property of the barrel be the same before and after the interchange.

For example, we could require that the weight of the barrel not change. A little thought will convince you that provided *every* apple is changed into an orange and vice versa, this requirement will always be met, no matter what the relative weights of the two pieces of fruit. We would say that the total weight is invariant under the substitution. Another example of this type of invariance occurs in strong interactions, where we say that the interaction is the same when we substitute neutrons for protons and vice versa. This is what we have called the requirement of isotopic spin invariance of the strong interactions.

A much more stringent set of requirements on the apples and oranges would arise from considering a different kind of operation. Suppose we required that the weight of the barrel not change when, as we encountered each piece of fruit, we chose randomly whether to make it into an apple or an orange.

Since now the type of fruit at each point in the barrel is the result of a random selection, the total weight will remain fixed only if the weight of the individual apples and oranges are the same. In this way, the requirement that the overall weight be invariant is equivalent to the requirement that individual weights be equal. Clearly, demanding invariance under this type of interchange is much more stringent than it was in our first example.

The hadronic analog of this random interchange would be to require that a theory predict the same values for measureable quantities if we interchange protons and neutrons in an arbitrary way at each point in space. Theories that satisfy this rather stringent requirement are said to be invariant under gauge transformations, and are called gauge theories. I use the plural here because there are many theories now in existence that incorporate this feature.

The gauge theories are attractive to physicists for a number of reasons. In the first place, they are aesthetically pleasing because they reduce the number of fundamental interactions from four to three. They also suggest lines of attack that might succeed in reducing this number even further, lines of attack that are now being vigorously pursued. They also manage to do away with some long-standing theoretical difficulties in the theory of weak interactions. In essence, gauge theories show that certain calculations which, in the old theory, gave infinite answers for certain cross sections, were incomplete because they did not include the effect of the neutral vector boson. Including this particle effectively cancels the terms that became infinite in the previous calculation. Technically speaking, theorists say that gauge theories make the weak interactions renormalizable. Finally, the gauge theories predict that there should be neutral current reactions, a prediction that has been verified explicitly by experiment.

Lest we leave with the impression that there is no connection at all between gauge theories and the idea of quarks, I turn to one of the gauge theories that is a "hot" item in present-day research. This is the hypothesis in which it is imagined that the interaction between quarks is to be described by a gauge theory. This is not motivated, as was the original Weinberg-Salam theory, by a desire to unify interactions. Indeed, there is no attempt to relate the strong interactions to anything else. It is simply an attempt, so far largely successful, to apply the features of the gauge theories to the problem of the hadrons.

The theory I refer to is called *quantum chromodynamics.* The "chromo" refers to the fact that it is a theory in which the constituents of the hadrons consist of quarks that have the color quantum number. The force between these quarks is mediated by a set of eight massive spin one particles called *gluons*—they "glue" the hadron together. The gluons have the property of being able to carry the color quantum number from one quark to another, much as charged pi-mesons in nuclei can carry electrical charge from one nucleon to another. In fact, if you think of the role of the pi-meson in the nucleus, you have a pretty good picture of the role of the gluons in elementary particles.

Quantum chromodynamics got quite a boost during the summer of 1979. A large group of researchers at the German Electron Synchrotron Laboratory in Hamburg, including Samuel Ting (who, you will recall, shared in the discovery of the ψ/J), announced the results of an experiment that gave strong evidence for the existence of gluons. In normal high-energy collisions, particles tend to be produced in narrow sprays, called jets. There is usually one jet moving in the direction of each of the two particles in the interaction. One way of understanding this result is to imagine each of the two colliding particles as being broken into pieces by the other. After having been split up, each of the collections of bits, which we perceive as a set of produced particles, continues to move along as it had before.

What the group at Hamburg found was that in a measureable fraction of very high-energy collisions of electrons and positrons, a third jet could be detected. This jet would be associated with the "shaking loose" of a gluon in the collision, and its subsequent decay into ordinary particles.

There have also been some attempts to formulate a theory that carries the Weinberg-Salam unification one step further and bring together the strong, electromagnetic, and weak interactions. These theories are called *grand unification schemes,* and are still highly speculative, but they do have one interesting feature. Because of the connection between strong and weak interactions, there is some probability that the proton will not be stable, but will undergo decay. These theories predict a half-life for the proton of about 10^{31} years—a lifetime that we saw in Chapter II is just about at the present experimental limit of observation.

Whether quantum chromodynamics and the grand unification schemes turn out to be correct or whether they join so many of their predecessors in the vast graveyard of interesting but incorrect theories, the principle of gauge invariance, which has brought about the unification of electromagnetic and weak interactions, has been an enormous advance in our search for simplicity in nature. For their work on guage theories, Weinberg, Salam, and Sheldon Glashow (of Harvard University) shared the Nobel Prize in 1979.

Unsolved Problems: Experimental

THE most important problems facing experimental high-energy physics at this juncture have to do with searches for particles that theorists claim ought to exist, but which have not yet been found. We have talked about two such particles in some detail, the quark and the intermediate vector boson. Since both these particles are extremely important in shaping our present ideas about nature, the rewards (both professional and personal) that would accrue to the person who finds them would be quite high. Hence, the intensive and extensive activity of the searches.

There are, however, other kinds of objects that are thought, for one reason or another, to be important enough to be searched for with some care. There follows a short list of these particles, but in going over the list one should keep reminding oneself that all of them are *hypothetical* particles, and that any (or all) of them could turn out to be nothing more than figments of a theorist's imagination.

Heavy Leptons. One of the important strains that runs through the recent theoretical work on the quark model is the idea that some sort of symmetry exists between quarks and leptons. Every time a new lepton is discovered, we expect to find a new pair of quarks. Hence, an obvious question is: "How many leptons are there?" Over the next decade we may expect a continuing search for these particles, as well as for the kinds of narrow resonances that indicate the presence of new quarks.

Magnetic Monopoles. The idea of the symmetrical pervades modern physics. Yet there is one glaring asymmetry, an asymmetry so obvious that most of us never even think about it. We know that electrical charge occurs in multiples of a fixed charge (the charge on the electron). Electricity and magnetism are closely related, and yet we cannot see anything in nature that is the magnetic analog of a single isolated electrical charge. Magnetic poles always occur in pairs—every magnet has both a north and a south pole. Theorists have speculated that an isolated north pole or south pole of a magnet could exist, and have named such an object the *magnetic monopole.*

If the monopole exists, then the laws of quantum mechanics suggest that it must have a very large strength—so large, in fact, that it would produce over 10,000 times as many ions in a bubble chamber as would a proton. This would make it relatively easy to detect. As I write, there is one claim that a monopole has been detected, the main evidence being a heavy track in an emulsion-type experiment that was flown in the Skylab. But in our uncertainty, as with the search for quarks, a ritual check for monopoles is made whenever a new accelerator is turned on.

We might mention in passing that if a monopole should exist, the fact that electrical charge comes only in units of the fundamental charge of the electron would follow from the basic laws of quantum mechanics and therefore would no longer be an unexplained regularity in nature.

Tachyons. There is a common piece of folklore that says that relativity requires that nothing can move faster than the speed of light. Actually, that is not quite true. What relativity requires is that no particle that is *now* moving at less than the speed of light can be accelerated up to (or beyond) the speed of light. There is no reason, however, why there could not be particles that are moving at speeds greater than the speed of light now, provided that they had always moved at those speeds in the past.

In other words, if you adopt this way of looking at things, the speed of light acts more as a curtain than as a barrier. On the other side of the curtain are particles that always move faster than light. These hypothetical particles are called *tachyons.* On our side of the curtain are particles that move more slowly than light—particles that are sometimes referred to as *tardyons.* All that relativity tells us is that the curtain cannot be crossed from either side.

Whether tachyons actually exist and, if so, what their properties would be, are open questions. If they should exist, however, and if they could be used for sending signals, we would have a means of virtually instantaneous communication with any point in the universe. I have always felt, for example, that the reason we have never picked up radio signals from extraterrestrial civilizations is that any *really* advanced technology would be signaling with tachyons, and would no more think of using radio than we would think of using smoke signals. So the search for tachyons is one aspect of high-energy physics that could have far-ranging philosophical consequences.

Proton Decay

THE grand unified field theories, because they view all forces as being equivalent in some sense, suggest an interesting possibility. What if there is a process by which the proton can decay into a positron and some mesons? This is not allowed for the strong interactions alone because a decay of this sort violates baryon conservation, but many of the unified theories predict a tiny probability for it to occur. The general theoretical consensus is that in these theories the predicted lifetime of the proton is 10^{31} years or so—tantalizingly close to the present limit of 10^{30} years discussed in Chapter II.

Stimulated by these predictions, several experiments are now being built to push the measured proton lifetime to from 10^{31} to 10^{32} years. The general technique is to put counters around a large tank of water and try to see the end products of the decay when an occasional proton actually goes through the interaction process. A positive result in such a search would be a tremendous boost for the concept of a unified field theory. The best guess is that results should be available in late 1981 or early 1982.

Unsolved Problems: Theoretical

AT present, the theoretical physics community is busily working out the consequences of the ideas of the gauge theories and trying to extend the unification principle to all four interactions. It will be several years before we will know whether these attempts will be successful or not. Meanwhile, the proliferation in quarks and leptons suggests that theories involving quarks might very well have to be replaced by theories involving "subquarks" before too long. This is a particularly unappealing prospect, but one that may soon be forced upon us, so that even if we do wind up with a unified field theory showing that the fundamental interactions are all identical at some level, we shall still be left with the problem of finding some underlying simplicity in the suddenly complex world of quarks.

These are what might therefore be called the near-term theoretical problems that have to be investigated. There are other, more fundamental questions that theoretical physicists have only begun to think about. These are the questions that ask why the fundamental constants of nature have the values that they have.

For example, one fundamental constant that we have used throughout the book is the charge of the electron. We have always treated this as a number that can be measured in an experiment, but which is a given as far as the theory is concerned. In other words, in discussing a theory that deals with particles that have electrical charge, we simply take the value of the charge from experiment.

It is clear, however, that the ultimate theory will not operate in this way. Such a theory will predict the charges and masses of the particles from first principles, in much the same way that the properties of geometrical figures are predicted by Euclid's postulates in plane geometry. In such a theory, even things such as Planck's constant will be predicted, rather than taken from experiment.

Of course, we are currently a long way from anything like an ultimate theory. One idea of how such a theory might come about was advanced in 1961 by Geoffrey F. Chew of Berkeley. It is called the *bootstrap* theory, and even though it has generally fallen into disuse, it has always been my favorite in the "ultimate theory sweepstakes."

The basic idea of the bootstrap is that there should be a small number of principles governing any theory of the behavior of particles. Some of the conservation laws we have discussed, for example, would be included among these select few. The only requirement that would be placed on this collection of first principles is that they be logically consistent; that is, that they not contradict one another. In its most radical form, the bootstrap hypothesis claims that there is one and only one set of principles that is logically consistent, so that they would comprise the only possible set of principles that could operate in the world. In answer to the question "Why is the world the way it is?" the "bootstrapper" replies that the world is the way it is because this is the only way the world can be and still be logically consistent. It is a powerful idea, but it is only one suggestion of the means by which an ultimate theory could arise. The difficulties in formulating the theory and in calculating predictions from it have, as I said, led to its eclipse in recent years. But it may some day make a comeback.

Epilogue

THROUGHOUT this book we have probed successively deeper into the "reality" of the physical world. The atom, the nucleus, and even the elementary particle have each given way to a deeper reality. At the present time, the deepest reality we have been able to reach—the quarks—are showing the same symptoms that we saw in each of the other "deepest" realities we have known. They are proliferating to the point where we can only wonder whether they constitute just another stage in the quest for understanding, rather than the end of that quest. Will the coming years show us that reality has even more layers to reveal, like some unimaginable cosmic onion? Or shall we find some way to end our quest at the level where we find ourselves now?

Only time will tell.

APPENDIX A

*NOBEL PRIZES AWARDED FOR WORK DISCUSSED IN THIS BOOK**

Year	Nobelist	Nationality	Work	Discussed in Chapter
1906	Joseph J. Thomson	British	discovery of the electron	I
1908	Ernest Rutherford	British	identification of radioactive emissions	I
1921	Albert Einstein	Swiss	photoelectric effect	I
1922	Niels Bohr	Danish	structure of the atom	I
1927	Charles T. R. Wilson	British	cloud chamber	IV
1932	Werner Heisenberg	German	quantum mechanics	III
1933	Paul A. M. Dirac	British	work in quantum mechanics	IV
	Erwin Schrödinger	Austrian		III
1935	James Chadwick	British	discovery of the neutron	II
1936	Carl D. Anderson	U.S.	discovery of the positron	IV
1939	Ernest O. Lawrence	U.S.	development of the cyclotron	VI
1945	Wolfgang Pauli	Austrian	exclusion principle	XII
1949	Hideki Yukawa	Japanese	prediction of mesons	III
1950	Cecil F. Powell	British	photographic emulsion techniques and the discovery of pi-mesons	V
1951	John D. Cockcroft	British	transmutation of the elements	VI
	Ernest T. S. Walton	Irish		
1957	Tsung-Dao Lee	U.S.	parity violation	XIII
	Chen Ning Yang	U.S.		
1958	Pavel A. Čerenkov	U.S.S.R.	Čerenkov counter	VII
1959	Owen Chamberlain	U.S.	discovery of the antiproton	VII
	Emilio Segrè	U.S.		
1960	Donald A. Glaser	U.S.	development of bubble chamber	VII
1961	Robert Hofstadter	U.S.	measurement of shape of proton and nuclei	VI
1968	Luis W. Alvarez	U.S.	study of unstable particles	VII
1969	Murray Gell-Mann	U.S.	theoretical physics	IX
1976	Burton Richter	U.S.	discovery of the Ψ/J	XII
	Samuel C. C. Ting	U.S.		
1979	Sheldon Glashow	U.S.	theory of weak interactions	XIV
	Abdus Salam			
	Steven Weinberg			

*This list does not include all Nobel Prizes awarded in particle and nuclear physics—only those discussed in this book.

APPENDIX B

SOME COMMON SYMBOLS

A—The number of nucleons in a nucleus
B—A magnetic field
c—The speed of light
h—Planck's constant
I—Isotopic spin
J—Spin
m_p—Mass of the proton
m_e—Mass of the electron
Q—Total charge of a particle
S—Strangeness
σ—Cross section
Z—Total charge of a nucleus

APPENDIX C

A CATALOG OF PARTICLES (Unstable)

Name	Symbol	Characteristics	Chapter Discussed
Mu-meson (muon)	μ	lepton	V
Pi-meson (pion)	π	meson	V
Lambda	Λ°	strange baryon	V
K-meson (kaon)	K	strange meson	V
Delta	Δ	nonstrange baryon	VII
Rho-meson	ρ	nonstrange meson	VII
Sigma	Σ	strange baryon	VII
Cascade	Ξ	doubly strange baryon	VII
Sigma 1,385	$\Sigma(1385)$	strange resonance	VII
Omega minus	Ω^-	triply strange baryon	X
Psi/J	Ψ/J	meson containing charmed quarks	XII
D-meson	D°	charmed meson	XII
Upsilon	Υ	meson containing bottom quark	XII
Tau-meson	τ	heavy lepton	XIII

A CATALOG OF PARTICLES (Stable)

Name	Symbol	Characteristics	Chapter Discussed
Electron	e	lepton	I
Photon	γ	light	I
Neutrino	ν	lepton	II, XIII
Proton	p	baryon	I
Neutron	n	baryon	II
Positron	e^+	antielectron	IV

In general, an antiparticle is denoted by a bar over the particle symbol. Thus, the antiproton is written $\bar{p}$ and so forth.

GLOSSARY

THE following terms are important in any discussion of elementary particles. In addition to a short definition, the chapter in which the term is introduced or discussed in the text is generally given in parentheses immediately after the term itself.

ATOM (I)—the "indivisible" smallest piece of matter first postulated by the Greeks, but now known to be composed of a nucleus circled by electrons.

BARYON (VIII)—any hadron that contains one proton in its final set of decay products.

BETA DECAY (II)—the process by which a neutron decays into a proton, an electron, and a neutrino. If the process occurs when the neutron is inside a nucleus, we speak of nuclear beta decay.

BOOTSTRAP (XIV)—a theory of elementary particles in which logical consistency is the ultimate requirement.

BOTTOM QUARK (XII)—one of the new quarks whose existence is shown by the discovery of the upsilon particle.

BUBBLE CHAMBER (VII)—a device in which the track of a particle crossing the chamber is marked by a string of bubbles condensing on ionized atoms.

C QUARK (XII)—the quark whose existence is demonstrated by the discovery of the ψ/J particle. It carries the charm quantum number.

CATHODE RAY (I)—an old term for electron.

ČERENKOV COUNTER (VII)—a device that identifies particles passing through it by observing a flash of light generated in a manner similar to a sonic boom.

CHARGE CONJUGATION (VIII)—the mathematical operation that turns a particle into its antiparticle.

CLOUD CHAMBER. See WILSON CLOUD CHAMBER.

COLOR (XII)—the property of quarks that allows them to be arranged in ways that seem to violate the Pauli principle.

CONSERVATION LAWS (XIII)—any observed regularity in nature which indicates that a particular quantity (electrical charge, for example) is the same before and after a reaction.

COSMIC RAYS (IV)—energetic particles (primarily protons) that are created in stars and enter the earth's atmosphere.

CYCLOTRON (VI)—a device that accelerates protons to high energies.

EIGHTFOLD WAY (IX)—a way of grouping the elementary particles that reveals regularities in their properties.

ELECTRON (I)—the small, negatively charged particle that normally circles the nucleus of an atom.

ELECTRON VOLT (abbreviated eV) (V). The energy acquired by one electron falling through one volt of potential energy.

ENERGY (IV)—the ability to do work. It comes in many forms (kinetic, potential, and mass) and is conserved.

FERMI (abbreviated as F) (I). A unit of length equal to 10^{-13} cm, about the distance across a proton.

FNAL (VI)—the Fermi National Accelerator Laboratory—the world's largest accelerator, located near Chicago.

FLAVOR (XII)—the aspect of a quark that tells which of the six kinds of quark it is.

GAUGE SYMMETRY (XIV)—a symmetry in which no measurable property of the world changes if protons and neutrons can be substituted for each other at each point in space independently.

GAUGE THEORY (XIII, XIV)—any theory that incorporates gauge symmetry.

GEIGER COUNTER (I)—a device that detects ions created by a charged particle.

GEOLOGICAL SEARCHES (XI)—searches for quarks presumably trapped in various materials on the earth.

GeV (V)—abbreviation for giga (10^9) electron volts of energy.

HADRON (VII)—any particle that participates in the strong interactions.

HALF-LIFE (II)—the time required for one-half of a given sample of particles or nuclei to decay.

HEAVY LEPTON (XIII)—a particle having properties similar to the electron or mu-meson, but more massive.

ION (I)—an atom from which one or more electrons have been stripped, leaving it with a net positive charge.

ISOTOPIC SPIN (VIII)—a mathematical quantity related to the number of different charges in a particle family.

keV (V)—abbreviation for kilo (1,000) electron volts of energy.

LEPTON (VII, XIII)—a particle (like the electron, muon, and neutrino) that participates in the weak, but not the strong, interactions.

LEPTON CONSERVATION (XIII)—a rule which states that the net number of leptons before and after an interaction must be the same.

LINEAR ACCELERATOR (VI)—an accelerator in which the particles move in a straight line as they gain energy.

MAGNETIC MONOPOLE (XIV)—a hypothetical particle that carries a single magnetic pole.

MAGNETIC SPECTROMETER (XII)—a device that uses the bending of particles in a magnetic field to separate out those of a given momentum.

MATHEMATICAL QUARKS (XI)—a term for quarks that might "exist" only in theory, but never be found in the laboratory.

MESON (V, VIII)—originally, any particle whose mass is between that of the electron and proton—in modern terms, any particle whose decay products do not include a baryon.

MeV (V)—abbreviation for mega (10^6) electron volts of energy.

MU-NEUTRINO (XIII)—the neutrino given off in the decay of the mu-meson.

NUCLEAR DEMOCRACY (VII)—the idea that every particle is equally "elementary."

NUCLEUS (I)—the heavy positively charged center of the atom, composed of protons and neutrons.

NUCLEON (II)—a term referring to both the proton and neutron.

NEUTRAL CURRENT (XIII)—the uncharged object exchanged when a neutrino scatters from a hadron.

NEUTRINO (II, XIII)—a zero-mass uncharged particle emitted in the process of beta decay.

NEUTRON (II)—a particle of approximately the same mass as the proton, but uncharged.

PARITY (VIII)—a mathematical operation which exchanges right and left.

PARITY VIOLATION (XIII)—refers to the observed fact that in some beta decay processes electrons are emitted preferentially in the right-hand direction.

PARTICLE-WAVE DUALITY (III)—a pseudodilemma brought about by the fact that elementary particles behave neither as waves nor as particles.

PAULI PRINCIPLE (XII)—the principle that states that two spin 1/2 particles such as electrons or quarks cannot occupy the same state.

PHOTON (I)—the "particle" associated with light.

POSITRON (IV)—the antiparticle of the electron.

PROTON (I)—the massive, positively charged particle that is the nucleus of the hydrogen atom.

QUARK (IX)—the hypothetical particle that is believed to be the basic constituent of the elementary particles.

QUARK CONFINEMENT (XI)—the theory that there is some reason why quarks might exist inside of elementary particles but may not be seen in any experiment.

PLANCK'S CONSTANT (I, III)—a fundamental constant of nature important in quantum mechanics.

SCINTILLATOR (I)—a type of particle detector that emits a flash of light when a particle strikes it.

SPIN (VIII)—the property of an elementary particle analogous to the rotation of the earth on its axis.

SLAC (VI)—acronym for Stanford Linear Accelerator Center, the world's largest electron accelerator.

STORAGE RINGS (VI, XII)—devices in which accelerated particles are kept moving in circles by magnetic fields.

STRANGENESS (V, VIII)—the property of elementary particles that governs the speed at which they decay.

STRONG FORCE (INTERACTION) (II, III)—the force or interaction responsible for holding the nucleus together.

SYNCHROTRON (VI)—an accelerator in which magnetic fields and acceleration are synchronized to keep the particles moving in a narrow ring.

TACHYONS (XIV)—hypothetical particles that move faster than the speed of light.

TeV—abbreviation for tera (10^{12}) electron volts of energy.

TIME REVERSAL (VIII)—the mathematical operation analogous to running a movie film backward.

TOP QUARK (XII)—the as yet undiscovered quark that is the partner of the bottom quark.

UNCERTAINTY PRINCIPLE (III)—the principle that states that it is impossible to measure both the position and momentum of a particle with infinite accuracy.

UNIFIED FIELD THEORY (XIV)—a theory in which two or more interactions are seen to be different aspects of a single process.

W BOSON (XII)—the hypothetical particle that is supposed to be exchanged in beta decay and other weak interactions.

WAVE FUNCTION (III)—in quantum mechanics, the mathematical function that gives the probability of finding the particle at a given point.

WEAK INTERACTIONS (II)—processes, like beta decay, that proceed slowly on the nuclear time scale.

WILSON CLOUD CHAMBER (IV)—a device used in early work that records the passage of charged particles by the presence of droplets formed on ions left by their passage.

Y* (IX)—an old expression for higher mass resonances associated with strange baryons.